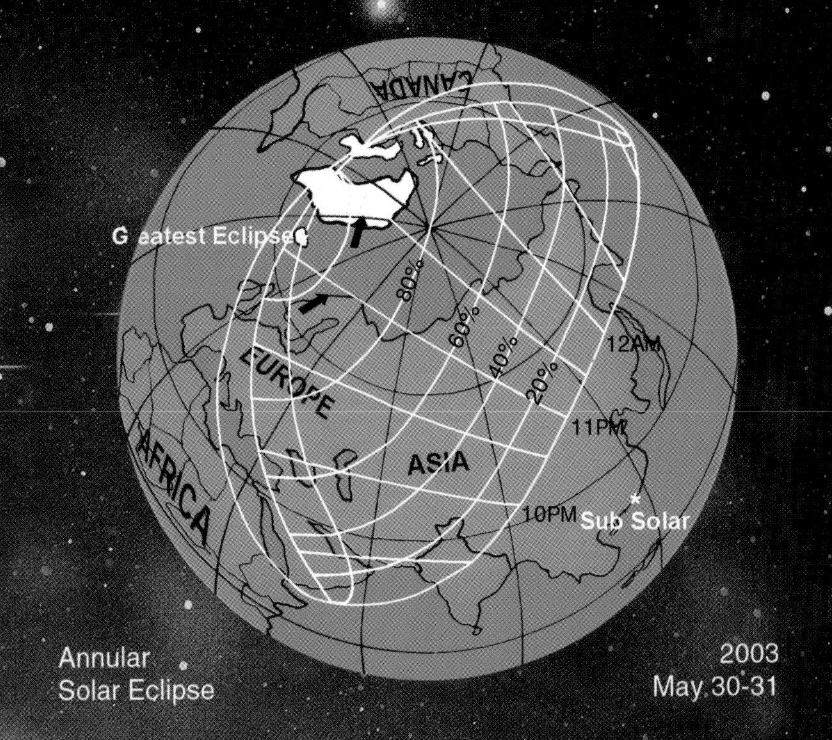

Annular
Solar Eclipse

2003
May 30-31

Total
Solar Eclipse

2008
August 1

What do the maps of solar eclipses on the preceding pages show?

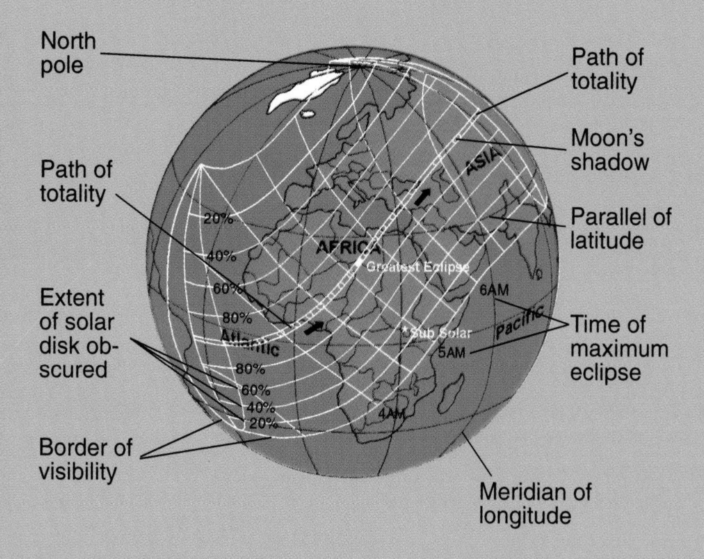

How Do I Use the Maps?

The maps on the foldout front flap show total and annular solar eclipses up to the year 2010 and include the exact date and time, as well as how much of the solar disk is obscured.

In a total solar eclipse, the Moon's shadow travels across Earth. The shape of the inner portion of the shadow, the umbra, depends on how it strikes the curved surface of Earth. Because Earth is round, the area covered by the umbra is sometimes elliptical. On both sides of the path of totality (double band) there is a large area that is reached only by the penumbra. Here the Sun is only partially darkened. The part of the Sun's disk that is covered grows larger the closer one gets to the location of totality. The maps indicate the time (Eastern Standard Time) when the umbra reaches a given point in the location of totality. For places reached by the penumbra, the time when the Sun is most obscured is given. The percentage numbers indicate the maximum obscuring of the solar disk by the Moon. For more on solar eclipses, turn to page 108.

Joachim Ekrutt

Stars and Planets

Second Edition

Identifying Them, Learning about Them, Experiencing Them

With all Important Celestial Events up to the Year 2010
and with a Lexicon of Celestial Bodies

175 Star Maps, Pictures of the Constellations, and Diagrams
30 Color Photos of Celestial Objects

Star Maps and Diagrams by Wil Tirion
Maps of Solar and Lunar Eclipses by Brian Sullivan

Consulting Editor: Clint Hatchett
Director
Science and Space Theatre
Pensacola, Florida

Preface/Contents

A clear, starry night, when the sky sparkles and glitters everywhere, arouses our sense of curiosity. What are the names of the stars that shine with such special brilliance? How can one tell the constellations in this confusing multitude of stars? What else is there to see in the sky? There are also spectacular celestial sights, such as solar and lunar eclipses, galaxies, and planets that no devoted stargazer wants to miss. Barron's *Stars and Planets* will keep anybody—whether beginner or experienced hobbyist—abreast of what is going on in the sky through the year 2010. Wil Tirion, an internationally acclaimed astronomical cartographer, has developed especially for this book 72 star maps that will help anybody get oriented in the sky at first try. The maps always show the section of the sky that one actually sees when looking north or south. Marked in these monthly maps are all the stars and constellations that are clearly distinguishable to the naked eye. And the maps can be used worldwide!

A special and novel feature in this book is its maps of the globe printed in color. They show at a glance when and where solar and lunar eclipses will be visible during the entire first decade of the twenty-first century. This is a feature that amateur astronomers (who do not want to miss these thrilling events) will appreciate. But that is not all: In an astronomical calendar all the most beautiful heavenly events—such as planets in easily viewable positions—are listed in easy-to-consult tables. The book also contains many helpful illustrations, accompanied by precise explanations, as well as a chapter on basic astronomy for beginners and a "Lexicon of Celestial Bodies," which serves as a useful reference and at the same time makes fascinating reading in its own right. All in all, this handbook of the decade is both a tool to help you identify what you see in the sky and a descriptive account that brings the heavens closer.

The author and editors of *Stars and Planets* wish you many happy and exciting hours of stargazing.

Contents

The Lagoon Nebula in Sagittarius, a glowing cloud of hydrogen gas (see page 142).

Star map N I/1, January, north.

Photo on following two pages: Saturn and its moons: a spectacular photo montage of pictures taken by the space probe Voyager I. ▷

The photos on the cover show:
Front cover: The star cluster of the Pleiades; the planet Saturn; a partial solar eclipse.
Back cover: Star map N I/1, January, south; Perseus from star map N I/2 north and Capricornus from star map S11 south. These illustrations are from Hevelius's sky atlas.
Spine of book: Lunar eclipse.

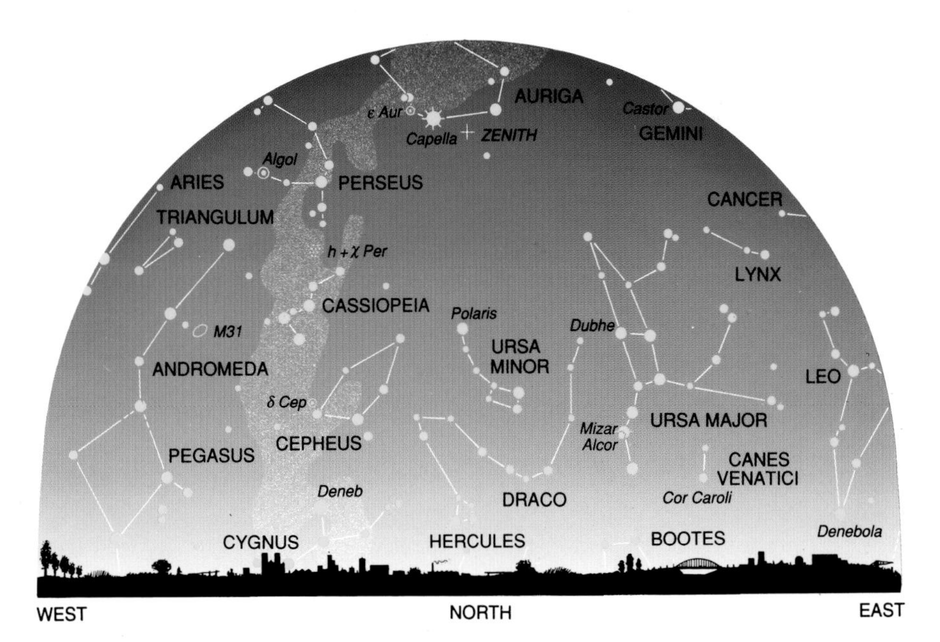

Astronomy Made Easy

Astronomy is one of the oldest sciences. Human beings have gazed up at the starry skies from time immemorial, and they recognized early on that strict laws govern the movement of the stars. The most important results of thousands of years of observation help us even today in studying and understanding the night skies.

The Movement of Earth

Our planet Earth is an almost perfect sphere with a diameter of 7,918 miles (12,756 km). It revolves around its own axis—an imaginary line that runs from the north pole to the south pole—with great precision. This rotation happens at an extremely fast rate. Thus, any point along the equator circles around the center of Earth at a speed of 1,036 miles (1,669 km) per hour. We are, however, completely unaware of this motion because everything rotates along with us—the air, the land, the oceans, our entire natural environment.

Earth takes 24 hours to revolve once. This is the basis of our measurement of time. Earth rotates from west to east. It would be extremely hard to prove that this motion actually takes place if it were not for the stars way beyond Earth's atmosphere. Unlike the surface of Earth, they remain still all day and all night. But because of Earth's revolving motion, they seem to travel across the sky in the opposite direction from Earth's rotation, namely from east to west. It follows that the rising in the east of stars, the Moon, the Sun, and all other heavenly bodies and their setting in the west are only illusions of movement produced by the rotation of Earth in the opposite direction.

Only very late, in the nineteenth century, were methods found that could prove without reliance on observation of the stars that Earth does indeed rotate around its axis. Probably the most famous of these methods is Foucault's pendulum, which is based on a brilliant experiment thought up by the French physicist Jean Bernard Léon Foucault. In 1851, Foucault had a pendulum hung in the dome of the famous Paris Pantheon. Because Earth rotates, the floor beneath the pendulum clearly shifted while the pendulum, on account of inertia, remained stationary. This was a sensational experiment that convincingly demonstrated to the fascinated public of the nineteenth century that Earth actually rotates and that it is not the stars that race around us at dizzying speeds, as people had so long imagined from ancient days through the Middle Ages.

How the Celestial Spheres and Earth Are Alike

Although we know today that there is no such thing as a celestial sphere—let alone one fashioned of crystal—from which the stars are somehow hung, this old way of picturing the world still has its uses. Let us imagine Earth, the terrestrial sphere, as a balloon that can be inflated until it touches this imaginary celestial sphere. If the surface of the inflated balloon Earth were marked with stamp ink, the most important lines on it would be transferred to the celestial sphere. Earth's equator (see diagram, page 9) would translate into the celestial equator, and the same would happen with the lines indicating longitude and latitude. The north geographic pole would mark the north celestial pole and the south geographic pole would yield the south celestial pole. This is exactly how astronomers look at the sky even today. The celestial equator divides the celestial sphere into a northern and a southern half, and the north and south celestial poles indicate the points in the celestial sphere that do not move along with the rotation but mark the axis around which the rotation takes place. Almost directly over the north celestial pole stands the famous North Star (see diagram, page 22). The south celestial pole is not marked by any particularly noticeable star (see page 22).

With the aid of this (imaginary) celestial sphere and the terrestrial sphere, it is easy to explain and picture the various movements of the heavenly bodies.

North Pole, South Pole, and Equator

An observer standing on Earth's north or south pole would find the celestial north or south pole directly overhead because Earth's axis is identical with that of the celestial sphere. For this observer, the pole is at the highest point of the visible sky, the zenith (see diagram below), and seen from here, all stars move along a path parallel to the horizon, which is the limit of the visible half of the heavens. The equator of the celestial sphere coincides exactly with the horizon. Hence only half the sky is visible throughout the entire year. Scientists at the American Scott-Amundsen south polar research station consequently see the same stars revolve continuously around them during the southern winter months.

The other extreme exists at the equator. Here the celestial equator rises up in an imaginary vertical line from the horizon and runs through the zenith, the highest point of the sky. The north celestial pole lies exactly on the horizon, as does the south celestial pole. An observer here can see all the stars of the sky. The stars in the east rise straight up in the east and sink straight down again below the horizon in the west.

For all geographical latitudes lying between these two extremes, there is a part of the sky that always remains invisible. It is the part surrounding the celestial pole of the opposite hemisphere. Thus, on Earth's

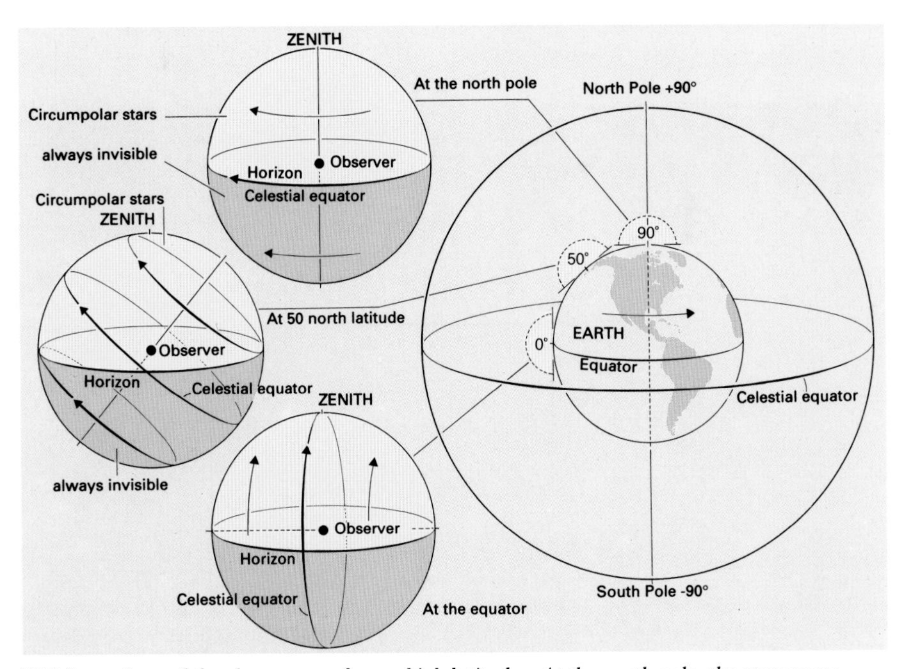

Which portions of the sky are seen from which latitudes: At the north pole, the stars move parallel to the horizon; at the equator, they move up and down vertically; in between (at about 50° north) they move at a diagonal.

Northern Hemisphere the area around the south celestial pole remains hidden. On the other hand, the area of the sky that is close to the visible pole remains in view all the time. It never "sets" but seems to circle around the celestial pole. The photo on page 129 shows this in a very striking fashion. These stars, which never set, are called circumpolar stars because they seem to revolve continually around the pole. The closer one is to the pole, the more circumpolar stars there are. This is because at the poles an entire celestial hemisphere seems to rotate parallel to the horizon and all the visible stars become circumpolar.

Note: For identifying the stars, only the latitude of the observation point matters.

Consequently a stargazer in Canada sees the same stars as someone watching the skies in central Europe or in northern Japan. Similarly, Buenos Aires, Cape Town, and Adelaide (Australia) all see the same star patterns overhead; in all three places the same stars are circumpolar and the same stars are invisible because they remain below the horizon all the time.

Fixed Stars Are Not As Immobile As They Seem

The earliest people who watched the starry skies several thousand years ago noticed that the stars rise and set but that apart from this motion they remain—with some remarkable exceptions—stationary in relation to one another. Sirius, for example, the brightest of all the stars, is always the same distance away from, say, Aldebaran or the twin stars Castor and Pollux. For this reason these stars were very early called "fixed" stars, that is, stars that seem to have their firm, fixed, and immovable place in the firmament. This was before people realized

A photo taken with long time exposure shows the stars as tracks of light created by Earth's rotation.

that it is Earth's spinning around its own axis that causes the apparent rising and setting of the stars. In those times people believed that all the celestial bodies were located on a large, crystal orb that revolved around Earth at dizzying speeds. Today we know, of course, that the stars are not immovable objects attached to a heavenly vault.

The Origin of the Constellations

The so-called fixed stars are such inconceivably large distances away from Earth that their motion would lead to a perceptible change of the star patterns only over an extremely long time span, perhaps over tens of thousands of years. The distances between stars are therefore measured not in miles but in light-years. One light-year, the distance traveled by a light ray in one year (at a speed of 186,000 miles or 300,000 km per second), is equal to 5.88 trillion miles (9.46 trillion km) that is, rounded off, a 6 followed by 12 zeroes. Even our very closest stellar neighbor, Alpha Centauri (see pages 130–31), is 4.3 light-years distant, and most other stars are considerably farther away. Viewed from such a literally astronomical distance, even the immense distances traveled by the stars shrink to angles too small to be measured. So today we still see the stars in practically the identical places where they appeared to the ancient Egyptians and Babylonians 4,000 to 5,000 years ago.

Given the apparent permanence of the stars' locations relative to each other, only a small step was required to arrive at the idea of constellations. Many of the earliest civilizations thought they recognized certain shapes in the patterns made up by the apparently immobile stars, and they imaginatively identified these shapes with figures from their myths and legends. This is how the constellations came into being, and they have become one of the most important symbols of the starry heavens. They are also of considerable help in getting oriented in the night sky, and they are still of great significance even in modern astronomy. What constellations are visible, where they are in relation to each other, and the stories behind their names are subjects covered in the texts accompanying the star maps for each month (starting on page 27).

The Sun—The Star of Our Solar System

There is one star whose rising and setting is of supreme importance to us, namely, the Sun. Because of Earth's rotation, the Sun, like the other stars, rises in the east, travels across the sky, and sets in the west. In the Southern Hemisphere, too—in Cape Town or in Melbourne, for instance—the Sun rises in the eastern sky. It reaches its zenith in the north, and then sinks below the horizon in the west. This rising and setting of the Sun is what gives us our bright days and dark nights and thus the most basic divisions of time determining our daily lives.

When the Sun rises above the horizon all the other stars near it fade because of the sun's intense brightness. Earth's atmosphere contributes significantly to the dominance of the Sun's light because it diffuses the Sun's rays so much across the entire visible celestial hemisphere that the light of no other heavenly bodies can compete with it—with the exception of the Moon and occasionally the planet Venus. Because of their considerable brightness, Venus and the Moon sometimes can be seen even during daylight. The stars, however, don't become visible until the Sun has disappeared below the western horizon.

There is no abrupt transition from brightness of day to darkness of night. Day and night are separated by a period of dusk, during which the sky gradually darkens in the east while sunlight still glows above the horizon for some time in the west. The stars don't appear until later, well after the Sun has set. Dusk, too, is caused by Earth's atmosphere, which continues to diffuse sunlight across the sky even after the Sun has sunk below the horizon.

Dusk, Midnight Sun, and Polar Night

Astronomers distinguish between different phases of dusk, phases that are always dependent on geographic latitude. This is because the Sun, like the stars, sets at varying angles to the horizon (see diagram below). At the equator and in the tropics in general—that is, at low geographic latitudes—dusk is very brief because the Sun travels toward the horizon along a vertical path and therefore drops deep below it very quickly. In the mid-latitudes dusk lasts considerably longer, and in the extreme northern and southern latitudes, finally, we enter regions where the Sun at certain periods of the year no longer descends low enough to allow for real darkness. These are the times of the famous white nights and the midnight sun, when the Sun no longer sets but becomes "circumpolar" itself. It never sets during the 24 hours of the day, that is, during Earth's entire daily rotation. Observers anywhere within one of the polar circles, which are located at 66.5 degrees north latitude and 66.5 degrees south latitude, respectively—places like Tromsö and Murmansk, for example—experience this unique phenomenon. For several weeks and even months the Sun never sets, and six months later it fails to rise for a similar period,

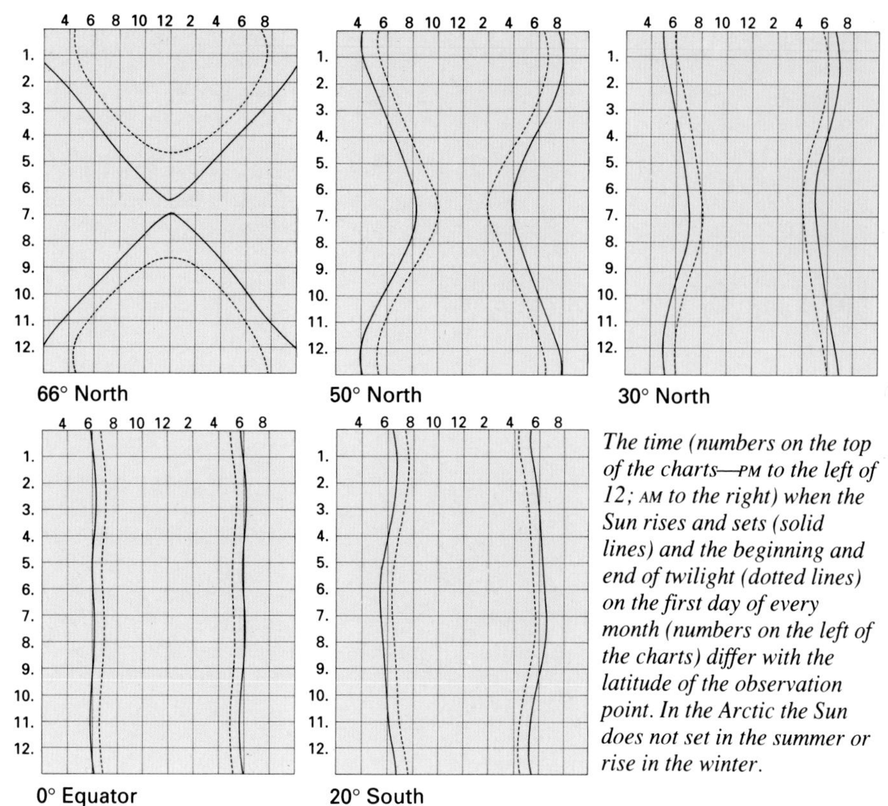

The time (numbers on the top of the charts—PM to the left of 12; AM to the right) when the Sun rises and sets (solid lines) and the beginning and end of twilight (dotted lines) on the first day of every month (numbers on the left of the charts) differ with the latitude of the observation point. In the Arctic the Sun does not set in the summer or rise in the winter.

creating the so-called polar night. The most extreme conditions are again found at the actual north and south poles, where the Sun is above the horizon for slightly over six months and remains below it for slightly less than six months.

All these phenomena—varying length of day, midnight sun, the polar night, as well as the seasons of the year—are the result of two facts: the movement of Earth around the Sun and the angle of Earth's axis in space (see diagram below).

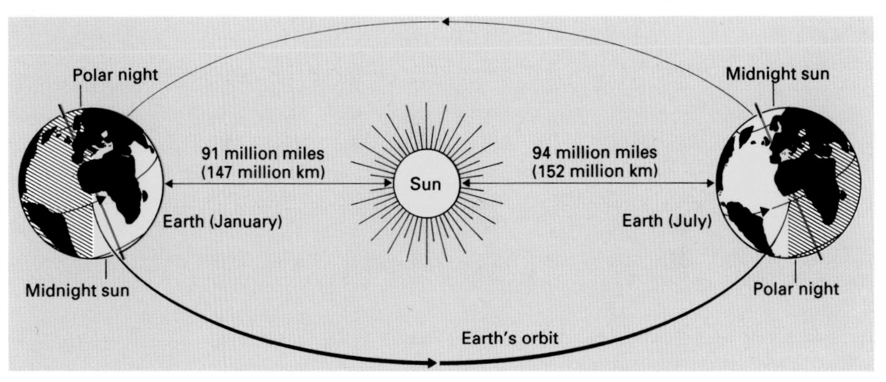

The position of Earth's axis (red line) is responsible for the seasons of the year.

What Causes the Seasons of the Year?

Earth not only revolves around its own axis but also moves around the Sun. The time Earth takes to complete one journey around the Sun (at an average velocity of about 18 miles or 30 km per second) is one year, which is the second time unit on which our measuring of time is based. To be exact, Earth takes 365 days, 5 hours, 48 minutes, and 46 seconds for one complete revolution around the Sun. In other words, Earth has rotated around its own axis 365¼ times by the time it has completed one circuit around the Sun. However, Earth's axis of rotation is not perpendicular to the plane of its orbit around the Sun but is instead at an angle of precisely 66½ degrees. This fact has significant consequences. Our seasons and all the phenomena described above are the result of this tilt of Earth's axis away from the perpendicular to its orbiting plane.

The above diagram illustrates the situation: At one time (in mid-December) Earth turns its southern half toward the Sun while the northern half faces away from the Sun. At this point the Sun is very high in the sky in the Southern Hemisphere and describes a large arc that leaves it shining much longer than 12 hours.

On the Northern Hemisphere, things are just the opposite: The Sun's rays emanate from low in the sky; they no longer reach the polar region, where the polar night reigns. In the Southern Hemisphere it is summer; in the Northern Hemisphere, winter.

Six months later (in mid-June), conditions are reversed. In between (in spring and autumn) the Sun shines down straight on the equator. For a short while, conditions are the same everywhere on Earth: The Sun shines for 12 hours everywhere, and the night also lasts 12 hours (on the seasons, see also page 104).

What Time Is It Where?

The rising and setting of the Sun also determines the time our clocks indicate. Our clocks imitate the heavenly clock, where the

Sun functions as a hand that indicates the time. We say it's 12 noon when the Sun reaches its highest point in the south (or, in the Southern Hemisphere, the north), and it's midnight when the Sun has sunk to its lowest point, invisible below the horizon. This model is much simplified. Our clocks indicate the time of day it is at a given meridian of longitude, and their rhythm is fine-tuned in accordance with an atomic clock. But even our modern time, which is defined by atomic clocks, still needs adjustment at irregular intervals to conform to the exact position of Earth in relation to the Sun.

All the times given in this book are in Eastern Standard Time (EST). EST corresponds exactly to the Sun's time on the 75th meridian of longitude west of Greenwich. To specify a location on the geographic globe, two coordinates are used: longitude and latitude. Latitude measures the distance from the equator both north and south. It is expressed in degrees or parallels of latitude, which, as the latter name suggests, run parallel to the equator. Longitude is the north-south coordinate. It also is measured in degrees, in this case counted east and west starting from the zero or prime meridian, which is the meridian that runs through the Royal Observatory in Greenwich, England. For purposes of practicality, the globe has been divided into time zones an average of 15 degrees wide, within which the same time, the so-called zone time, holds. The time difference from one zone to the next is one hour.

In addition to Eastern Standard Time, there are many other time zones, the most important of which is Greenwich Mean Time (GMT) and Universal Time because many astronomical events are referred to in this time. Today Great Britain and Portugal are among the countries that go by GMT. Spain, France, Germany, and Italy, among other countries, use Central European Time (CET). East of the CET zone one shifts to East European Time, which is in force in Romania, Greece, and Turkey among other countries. With the aid of the world time zone map below it is easy to translate EST or GMT into other local times. For places east of Europe, add the number of zones between 0 (Greenwich) and your zone to the local time; for places west of Europe, subtract the

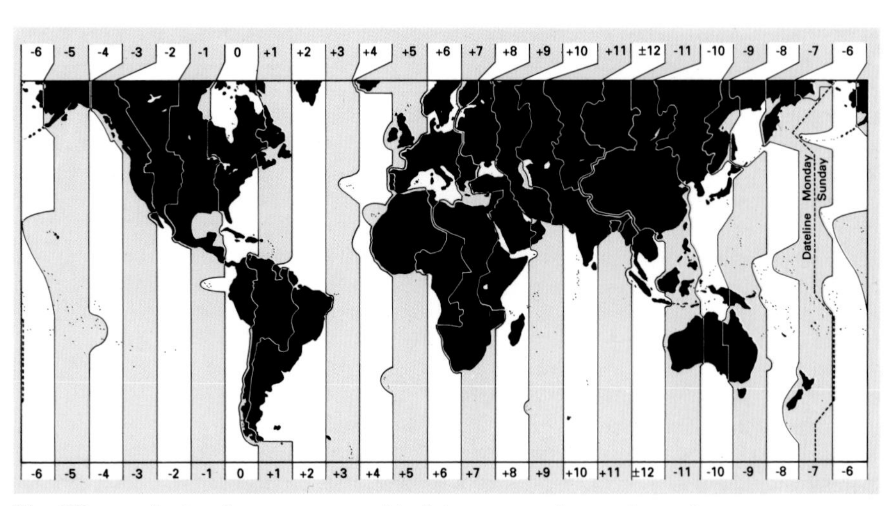

The difference in time from one geographical time zone to the next is one hour.

number. Thus, for example, the beginning of spring for the year 2005 is given on page 104 as March 20 at 7:32 A.M. For someone living in Germany this would mean that spring starts at 1:32 P.M. local time, whereas the clocks in Tokyo would say 21:32 or 9:32 P.M.

Many countries observe daylight saving time during the summer. This is done to save energy, as the name implies. For half the year (in the United States it is always from the first Sunday in April to the last Sunday in October) the clocks are set one hour ahead of standard time, that is, they indicate the Sun's time of the zone one slot to the east. All the heavenly bodies then rise one hour later according to the clock, and the Sun shines less brightly on the still sleeping cities in the morning and sets one hour later. Evening, thus artificially postponed, lasts one hour longer. When daylight saving time is in effect, all indicated times have to have one hour added to them. This turns EST into EDT (Eastern Daylight Saving Time). When using the world time zone map, one has to be careful because other countries also institute daylight saving time, but its period may not coincide exactly with that of EDT. Only if the United States and the country you are interested in have daylight saving time for the same period can you use a given star map without adaptation. Otherwise there will be a difference of minus one hour (for places east) or plus one hour (for places west).

Note: If you want to find out where and for how long daylight saving time is in effect and what time zone applies to a given place, the flight schedules of air lines offer exact information.

Everything About the Apparent Path of the Sun

Just as for a long time—until experiments such as Foucault's pendulum were invented—Earth's rotation could only be deduced indirectly from the apparent move-

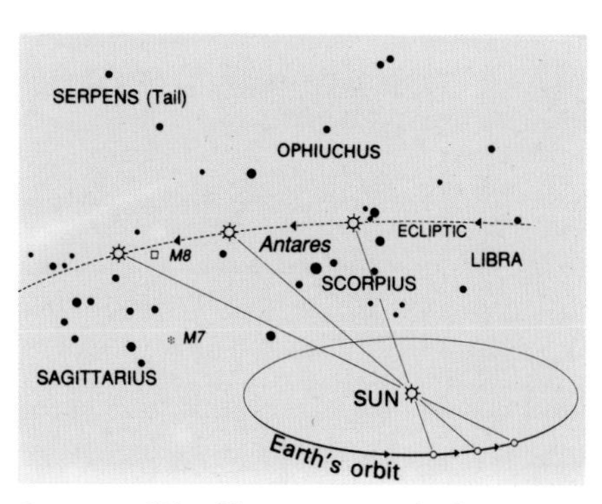

Because of Earth's movement, the Sun seems to travel from star to star.

ment of the firmament, the rising and descending of the stars, so Earth's movement through space could be recognized only indirectly by studying the movement of the Sun or, more accurately, its apparent movement. The Sun is much closer to Earth than the fixed stars. The latter are many hundred thousand times farther away from us. A glance at the diagram at the top of the page shows that from different points on Earth's orbit we see the Sun standing *in front of* different stars in the sky. The Sun seems to move and appears sometimes in front of more northerly constellations and at other times in front of more southerly ones. This apparent circular path the Sun takes across the sky in the course of a year is called the *ecliptic*. The Sun takes exactly one year for its journey around the entire celestial sphere, the same time Earth takes to orbit around the Sun. The Sun's apparent movement across the sky in front of the stars always follows the same path, the ecliptic. This word, which is derived from the Greek word *ekleipsis*, which means to omit or fail, is obviously related to *eclipse*. The imaginary path of the Sun is named the ecliptic because the Sun and Moon are always on the line of the ecliptic during solar and lunar eclipses (see pages 108 and 112). Astronomers of ancient Babylon already recognized this fact several thousand years before Christ. The ecliptic is one of the most important lines in the sky, and it is drawn into all the star maps as a broken line.

The Constellations and the Zodiac

Twelve or, to be more precise, 13 constellations lie along the ecliptic, and the Sun appears to pass through only these constellations in the course of a year. They are the famous constellations of the zodiac, which are among the very oldest constellations. They are:

The Constellations of the Zodiac

Aries the Ram (April 18)	Scorpius the Scorpion (November 22)
Taurus the Bull (May 13)	Ophiuchus the Serpent Bearer (November 29)
Gemini the Twins (June 21)	Sagittarius the Archer (December 18)
Cancer the Crab (July 20)	Capricornus the Goat (January 19)
Leo the Lion (August 10)	Aquarius the Water Bearer (February 16)
Virgo the Virgin (September 16)	Pisces the Fishes (March 11)
Libra the Balance (October 30)	

The dates in parentheses indicate the days on which the Sun enters the various constellations.

These dates bear no relation to those that are used in horoscopes.

The name zodiac is not altogether accurate because the constellations do not all represent animals but also include some human figures (Gemini, Virgo, Sagittarius, and Aquarius) as well as an object (Libra). Ophiuchus the Serpent Bearer is a completely different case. It did not figure in the classical zodiac at all. It is only since 1930, when the boundaries of all the constellations were revised, that Ophiuchus, which extends into the ecliptic belt next to Scorpius, has been counted among the zodiacal constellations (see pages 66–67).

All stars in the immediate vicinity of the bright Sun are invisible to us because they traverse the daytime sky along with the Sun. It is the stars on the opposite side of the heavens at any given time that we see in the night sky. When the Sun, on its apparent journey across the sky, reaches these stars half a year later, then those we saw earlier become invisible. Now the stars we couldn't see earlier lie opposite the Sun and can be seen at night. Thus the Sun determines which part of the starry sky we see at different times of the year.

The Moon and the Planets

The apparent path of the Sun, the ecliptic, is of great importance in orienting oneself in the night sky. For the Moon and the planets move along the same line, that is, they are always seen only within the 13 zodiacal constellations. Thus all the conspicuous celestial bodies move near the ecliptic.

Apart from the Sun, the Moon is the most prominent object in the sky. Except for the Sun, it is the only celestial object in which we can distinguish surface details with the unaided eye. There are lighter and darker spots that taken together and with a little bit of imagination can be seen as the famous "man in the moon" (for more on the Moon as a celestial body, see page 141). The Moon passes by the stars of the zodiac once every month and is never more than 5 degrees north or south of the ecliptic. During its orbit it goes through its famous phases, appearing in the sky now as half-moon, now as full moon, and, at new moon, disappearing from view altogether. Several times a year it causes the most striking celestial events of all, solar and lunar eclipses. Eclipses and the phases of the Moon are among the most dramatic variable events that take place in the sky, and they therefore are described in detail starting on page 105.

Planets are heavenly bodies that revolve around the Sun like Earth. Our Sun has nine planets that are up to 3.7 billion miles (6 billion km) away from it and orbit around it at various speeds. Together with innumerable smaller bodies—the comets (see page 133), meteorites (see page 138), and minor planets (see page 143)—they make up the solar system. All the bodies of the solar system are lit up by the Sun's light, and it is this light that they reflect toward Earth. In contrast to this, the fixed stars and all other heavenly bodies, such as galaxies, nebulas, and star clusters (see pages 135, 141, and 147), generate their own light.

The planets' order from the Sun is easy to remember with the aid of this mnemonic sentence in which the first letters of the words are the initials of planets' names: **M**y **V**ery **E**agle-eyed **M**other **J**ust **S**potted **U**mpteen **N**ew **P**lanets.

Five planets can be seen without the aid of a telescope or binoculars. They are the most luminous celestial bodies after the Sun and the Moon and almost always outshine even the brightest stars. The planets move along the ecliptic from one zodiacal constellation to the next. This moving or "wandering" results from the planets' revolving around the Sun, a fact reflected in the word planet, which is derived from the Greek word *planasthai*, to wander. The precise positions of the planets during the coming years are described starting on page 113.

The nine major planets, as well as Halley's Comet, orbit around the Sun.
The planets Saturn, Jupiter, and Mars as seen from Earth through a telescope.

Identifying and Experiencing the Stars

You need some theoretical background plus a bit of practical know-how to find your way around the night sky. But it is really quite amazing how little is required to get one's bearings in the vastness of the starry heavens. Start out by learning everything necessary for using the star maps that follow.

What Do the Star Maps Include?

The maps starting on page 27 include all the stars down to what is called the fourth magnitude, that is, all the stars easily visible to the naked eye. Fainter stars that barely can be discerned without magnification actually are visible these days only far away from cities and other lights and only under the best atmospheric conditions. The size of the yellow dot marking a star shows the star's brightness. The larger the dot, the brighter the star. The ten brightest stars—those that remain visible even against the artificially illuminated night sky of big cities—are indicated by small rays emanating from the dots. They are discussed separately in the lexicon section at the end of the book starting on page 130.

The solid lines linking groups of stars together indicate the constellations. Today 88 constellations, with boundaries defined by a commission of the International Astronomical Union in 1930, are recognized. They extend across the entire sky leaving no areas unaccounted for. Some of these constellations are made up of stars fainter than the fourth magnitude, and they therefore are not mentioned in this book. The star maps depict 62 constellations, all of which are clearly visible. The lines connecting groups of stars indicate especially striking configurations that formed the basis of the stories associated with the constellations and are thus directly responsible for their names. Every view of the sky is described in detail in the text below the maps.

◁ *The Big Dipper shining over the Calar Alto observatory in Spain.*

Which Star Map Applies Where?

The star maps are organized in three series, one each for three latitudinal zones, namely:

- 40 to 60 degrees north (NI)
- 20 to 40 degrees north (NII)
- 10 to 30 degrees south (S)

The first thing to do, therefore, is to determine in which of these zones your observation point is located. The geographic map preceding each series helps you do this.

The next step is to figure out the time when you wish to observe the sky. All the star maps reflect the sky as it appears at exactly the beginning of the month at 11 P.M., the middle of the month at 10 P.M., and the end of the month at 9 P.M.

If daylight saving time is in effect in a given location, an hour has to be added to these times, making them midnight for the beginning, 11 P.M. for the middle, and 10 P.M. for the end of the month.

No adjustments need to be made for the different time zones shown in the map on page 14. The times given are local times that apply to all the zones. Whether you watch the sky in San Francisco at 11 P.M. Pacific Standard Time, or at 11 P.M. Central European Time in Austria, the same chart applies. In some time zones the time difference between the western and eastern boundaries is more than one hour, but this inaccuracy is of no practical significance given the small scale of the maps. Only daylight saving time has to be taken into account because it throws the validity of all of the maps off by one hour. But there is a note to this effect on the maps where this may be the case.

If you want to watch the stars at some other hour, you will have to refer to the table on page 20. First find the month at the top, then the desired time of night at the left. Where the column from the top and the line from the left intersect there is a number that indicates the star map to be used.

Here are two examples:

Let's say you would like to watch the stars in Italy (the world map shows that you have to use the series NII) at 1 A.M. on July 20. In July it is daylight saving time, which means that you have to subtract one hour from what your watch says, that is, you have to find the line indicating midnight. At the intersection of the column for mid-July and the horizontal bar for 11 to 12 P.M. you find the number 8. Consequently you should turn to chart NII/8 on page 68.

Or: You are in South Africa and want to know if the Southern Cross is visible on December 28 at 3:30 A.M. December 28 can be considered the beginning of January, and where the two appropriate lines intersect you find—nothing! The missing part of the sky lies between what is shown on maps 3 and 4. Turn to map S3 (South Africa!) on page 85 and, if the star you are looking for is close to the horizon, also consult S4 on page 87.

Don't worry about the fact that you are using maps labeled August (in the first example) and March (in the second) even though you are observing in July and December, respectively. The months given apply only for the usual stargazing time from 9 to 11 P.M. (or 10 to 12 P.M. daylight saving time).

How Do I Find the Right Star Map?

	Jan.		Feb.		Mar.		April		May		June		July		Aug.		Sept.		Oct.		Nov.		Dec.	
	B.	M.	B.	M.	B.	M.	B.	M.	B.	M.	B.	M.	B.	M.	B.	M.	B.	M.	B.	M.	B.	M.	B.	M.
6–7 PM	11		12		1		2		3		4		5		6		7		8		9		10	
7–8 PM		12		1		2		3		4		5		6		7		8		9		10		11
8–9 PM	12		1		2		3		4		5		6		7		8		9		10		11	
9–10 PM		1		2		3		4		5		6		7		8		9		10		11		12
10–11 PM	1		2		3		4		5		6		7		8		9		10		11		12	
11–12 PM		2		3		4		5		6		7		8		9		10		11		12		1
12–1 AM	2		3		4		5		6		7		8		9		10		11		12		1	
1–2 AM		3		4		5		6		7		8		9		10		11		12		1		2
2–3 AM	3		4		5		6		7		8		9		10		11		12		1		2	
3–4 AM		4		5		6		7		8		9		10		11		12		1		2		3
4–5 AM	4		5		6		7		8		9		10		11		12		1		2		3	
5–6 AM		5		6		7		8		9		10		11		12		1		2		3		4

Important note: B = Beginning of month; M = Middle of month. Numerals in the columns under B and M designate the month of the correct map: 1 = January, 2 = February, 3 = March, etc. When daylight saving time is in effect (normally from April to October), one hour has to be added to the times given in the table. Perhaps a simpler method is to subtract one hour from the time, according to your watch, when you plan to observe the stars and use this time when consulting the above table.

The Constellations and Their History

To illustrate some of the old myths associated with the constellations, details from a famous sky atlas created by the Danzig astronomer Johannes Hevelius (1611–1687) have been reproduced. In this atlas, figures from Greek mythology are depicted as well as ones of more recent origin. Only in a few cases can one readily see how a star pattern inspired the name of its constellation. If you compare the figures from the Hevelius atlas with the stars in the sky, you will notice that the former show a mirror image of what we see overhead. Hevelius depicted the heavenly sphere as seen from the outside, so to speak.

Especially bright stars have proper names of their own. The names of stars like Sirius, Capella, and Aldebaran have been passed down from generation to generation and are, like the figures of the constellations, of ancient origin. Whereas most of today's constellations go back to Greek mythology, the stars were named by Arabian astronomers in the period between A.D. 600 and 1000.

Fainter stars are designated according to a system devised by Johann Bayer (1572–1625), a German lawyer and astronomer who published a famous sky atlas, the *Uranometria*, in 1603 in Augsburg. In this atlas he not only introduced many new constellations that are still recognized today but also invented a very practical system for naming stars, based on the Greek alphabet. The brightest star is named alpha (α), the second brightest, beta (β), and so on. Added to the Greek letter is the genitive of the constellation's Latin name.

Star Designations in the Star Maps
α Cen = Alpha in Centaurus
δ Cep = Delta in Cepheus
ε Lyr = Epsilon in Lyra
ε Aur = Epsilon in Auriga
η Car = Eta in Carina
χ Per = Chi in Perseus
ω Cen = Omega in Centaurus

Later on, astronomers also used regular Latin letters, as, for example, in h Per (= h in Perseus) or L_2 Pup (= L_2 in Puppis).

Symbols Used in the Star Maps

The star maps also show other celestial objects, which are represented by the following symbols:

Multiple stars: Here two or more stars revolve around each other in space, but to the unaided eye they appear as two stars that are exceptionally close to each other (see Binaries, page 131).

◉ **Variable stars:** In contrast to most other stars, the brightness of variable stars fluctuates (see pages 150–51).

□ **Nebulas:** Between the stars there are immense gas and dust clouds, some of which appear in the most amazing colors and shapes and count among the most beautiful objects of the night sky (see pages 141–42).

∴ **Star clusters:** The stars are not distributed evenly across the sky but often are concentrated in certain spots. Such aggregations of stars are called clusters. Astronomers distinguish between open and globular star clusters (see page 147).

O **Galaxies:** Galaxies are large star systems distributed throughout the universe. The distances that separate them from Earth—millions of light-years—are so immense that our imagination fails to grasp them (see page 135).

Nebulas, star clusters, and galaxies are indicated on the star maps by the letter M followed by a number; for instance, M 31 or M 8. M stands for Charles Messier, a French astronomer (1730–1817) who in 1771 published a catalog of celestial objects that appear cloudlike through a telescope. His catalog designations still are in use today.

All these objects are described in the text below the relevant star maps (pages 27 and following) and/or in the lexicon section starting on page 130.

Finally, the charts also show the night sky's most unusual feature, the luminous band of the Milky Way, which is depicted in the maps as a whitish area. The Milky Way is made up of the combined light of many millions of stars that are not individually visible but whose collective brightness is clearly apparent. The Milky Way snakes across the entire starry sky and in some places breaks up into branches. Its course and some especially interesting aspects of it are described in the text accompanying the appropriate maps and, under its own heading, on page 140.

Which Stars Do You Look for First?

Once you have determined which map to consult, you have to get oriented, that is, you have to figure out approximately where geographical north and south are. Here are a few pointers: In the Northern Hemisphere, Polaris, or the North Star, is a safe guidepost because for all practical purposes it is identical with the north celestial pole. If you draw a vertical line down from it with your eyes to the horizon, you are facing north. The North Star—and consequently the north celestial pole—is located most easily by first finding the best known star pattern, the Big Dipper (part of the constellation Ursa Major), which is circumpolar down to almost the middle of the geographic zone N II and therefore never disappears below the horizon in those northern latitudes.

Note: If the imaginary line connecting the two stars that form the ladle's edge away from the handle is extended, it leads directly to the north celestial pole. This way you will know where north is (see drawing below, left).

In the Southern Hemisphere, the Southern Cross will help you get oriented relatively easily. If the central axis of the Cross is extended, it points more or less accurately to the south celestial pole. Directly underneath the pole is geographic south (see drawing below, right).

Once you have established the directions in the sky, you are ready for further exploration. By comparing the map with the sky, you will find it relatively easy to locate individual constellations and stars. It is best to

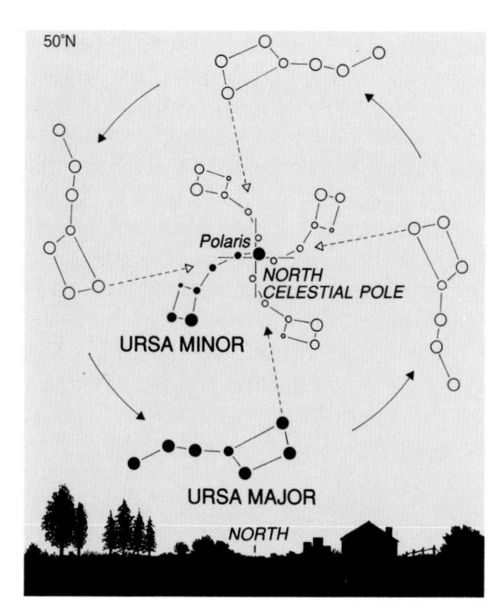

The two stars at the front of the bowl of the dipper point at the north celestial pole.

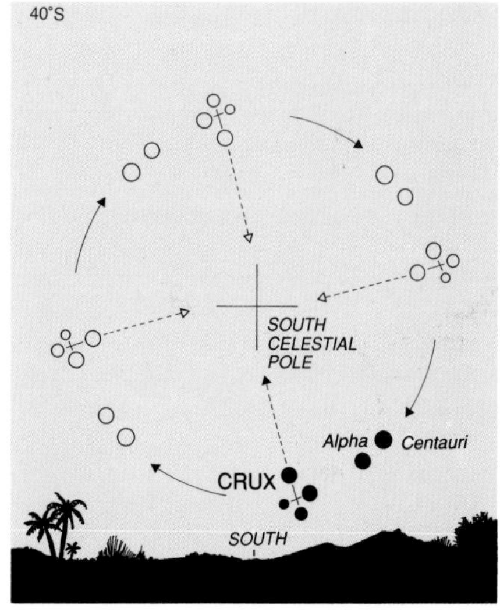

The central axis of the Southern Cross points at the south celestial pole.

use a dimmed flashlight for reading the maps. If possible, you should do your stargazing when the horizon is dark and there is little interference from artificial light. This is, of course, practically impossible in today's cities, where the sky is usually so bright that one can make out only the most brilliant stars.

Because of this, the very brightest stars—those that are visible even in the lit up skies above big cities—are marked in the maps with little rays. Start out by finding these stars first. Then see if you can find less radiant stars, which are indicated in the charts by smaller dots.

When exploring the sky, it is best to look for constellations. The thin connecting lines in the maps show the approximate outlines of the constellations as they actually are seen in the sky. Start with the brightest visible star and try to memorize the pattern formed by the nearby stars. If this pattern conforms to the picture in the map, you can go on and identify fainter stars and constellations. Without the help of the prominent constellation patterns, it would be a lot harder finding one's way around in the sky.

Below each sky map the most noticeable stars and constellations are described. Also noted are exceptional celestial events of the past. However, the brightest objects in the sky, the planets, do not appear in the maps at all because they move too fast. Therefore, if you spot an especially bright point of light in the sky that you cannot find in the map, check first to see if it is in the ecliptic. If it is—that is, if it appears in the constellations of the zodiac—you can turn to pages 113 and following to check which planets are visible in a particular year and month.

Practical Tips on Identifying Stars

It is always hard to get oriented in the sky when there are only a few stars out. If that is the case, it is better to wait a while, until you begin to recognize some of the constellations, that is, until the positions of individual stars in relation to each other become apparent.

In the north, Ursa Major, the Great Bear, with its well-known Big Dipper (called the Plough in England), is of great help in getting started. Once you have identified it, you know where north is and then it is easy to find other constellations (see drawing on page 22).

When dusk has just fallen and the western sky is still lit up by the setting Sun, all you can see are the brightest planets. A brilliant dot in the western sky is then almost always Venus, the evening star. When dawn begins to brighten the sky shortly before sunrise, what you see is usually again Venus, now figuring as the morning star (see page 151).

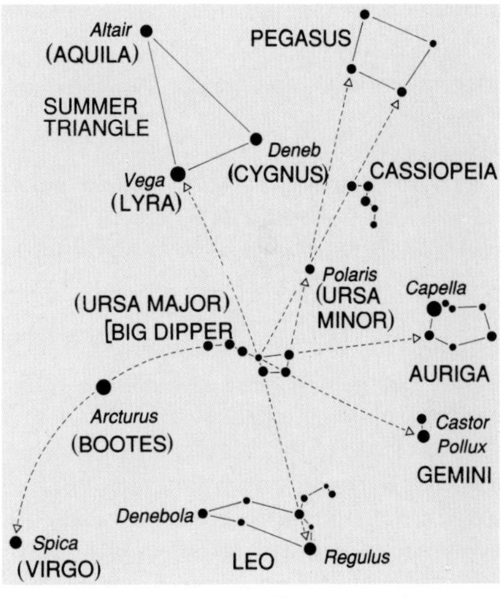

The Big Dipper as a guide post to other stars.

Note: The light of the planets does not flicker but is almost always steady. The stars, by contrast, often flicker or twinkle very noticeably, especially near the horizon. This is caused by turbulences in Earth's atmosphere. The steady light of the planets is one

A dust cloud in the constellation Orion is called the Horsehead Nebula due to its shape. A particularly bright star shines in the left half of the picture.

of the major signs that help distinguish them from the stars.

Before you go out to watch the stars, always check which planets are in the sky during a given month. You will find this information on pages 114 to 128.

Watching the Sky Through Binoculars

All the stars, planets, and other celestial phenomena mentioned in this book are clearly visible to the unaided eye. Celestial bodies that can be seen only through a telescope are not described. But because almost everyone owns a pair of binoculars, a few especially interesting celestial objects that are clearly visible through binoculars are mentioned.

Any ordinary field glasses will do for watching the sky, but they should be held motionless. This is easily done with the help of a photographer's tripod and a small clamp that is attached to the central part of the binoculars. If you try to watch the sky through handheld binoculars, the normal tremor of your hands is too great, no matter how hard you try to control it, to give you a picture of the stars that does not move. But the use of binoculars mounted on a tripod will open up unexpected possibilities. The increase in the number of stars and other heavenly bodies seen through binoculars versus what the naked-eye observer sees is greater than that yielded by replacing the binoculars with an average-sized telescope.

The Moon is the most important heavenly body on which even small binoculars reveal quite a few features.

Note: The light of the Sun is too brilliant to look into directly, let alone through binoculars. If you looked at the Sun without eye protection, your retinas would be destroyed instantly.

Although most of the heavenly objects described in this book can be viewed with the naked eye, it is highly recommended

The galaxy M 51 in the constellation Canes Venatici (the Hunting Dogs) is 20 million light-years away.

The star cluster the Praesepe, or Beehive, in the constellation Cancer (see page 145).

looking at them at least once with binoculars—except for the Sun, of course. There is just one fact you should be aware of to prevent disappointment: Colors are not discernable, whether you look with the naked eye or through binoculars. Only a few unusually brilliant objects show traces of color; Mars and Betelgeuse, for instance, are slightly reddish; Jupiter looks yellowish, and Sirius is white. All other celestial objects, especially the far away nebulas and galaxies, show up only as tiny, dimly shimmering, unimpressive blurs, even through binoculars. The famous color photos taken at the great observatories were photographed through huge telescopes, sometimes with hour-long exposures, and then underwent complicated photographic processes. Although an attempt is made to show (as well as possible through the Earth's distorting atmosphere) the proper colors, these pictures do not transmit the colors our eyes can perceive directly but reflect various frequencies or wavelengths the objects emit. Long wavelengths appear red, short ones, blue. If you keep all this in mind, you are not likely to experience disappointment when you observe the fascinating night sky.

Star Map Series N I

The map series N I shows the night sky as seen in all countries and cities lying in the latitude between 40 and 60 degrees north.

The farther south you are in this zone, the higher the stars will appear in the southern sky and the lower in the north. Vice versa, the stars will sink lower in the south and be higher in the north as you approach the zone's northerly end at 60 degrees north. The position of the stars relative to each other does not, of course, change.

All the star maps show the night sky as it looks at 11 P.M. at the beginning of the month, at 10 P.M. in the middle of the month, and at 9 P.M. at the end of the month. If daylight saving time is in effect, these times change to midnight, 11 P.M., and 10 P.M., respectively. If you want to watch the sky at another hour, you have to consult the table on page 20.

During the summer months, it does not get dark enough in the countries north of the 55th parallel of latitude for all the stars to become visible. North of the polar circle the Sun is in the sky 24 hours a day.

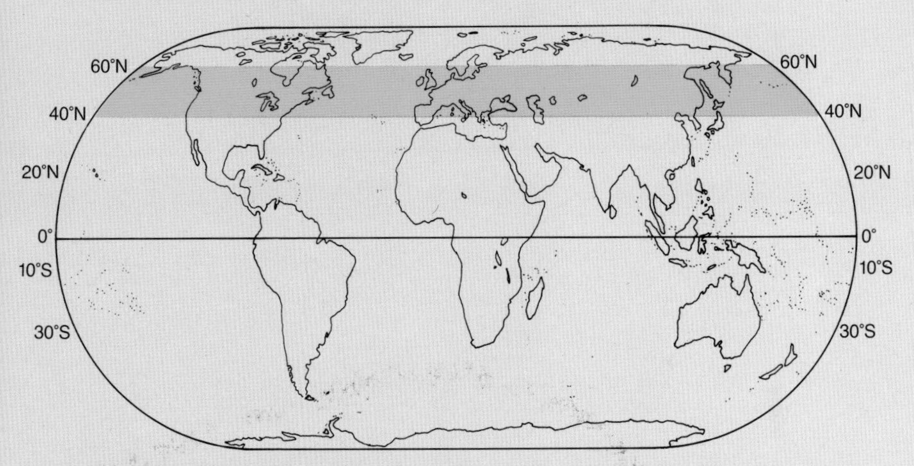

The star maps N I apply to the following areas and countries: all the northern states of the United States, Canada, Scandinavia, Germany, Belgium, Holland, Luxembourg, Switzerland, northern Italy, France, Great Britain, Poland, The Czech Republic, Hungary, Romania, Yugoslavia, Bulgaria, and the entire former Soviet Union.

What the symbols in the star maps stand for:

- ● = Double stars: Pairs of stars that are particularly close to each other.
- ◉ = Variable stars: Stars that change in brightness.
- □ = Nebulas: Clouds of gas and dust that shine in colors.
- ∵ = Star clusters: Aggregations of many stars concentrated in one place.
- O = Galaxies: Star systems (the Milky Way is one of many galaxies).

About the brightness and size of the stars marked in the maps:

0 = ✸ The brightest stars. They are easily visible even in big city skies.

1 = ● 2 = ● 3 = ● 4 = • Scale of brightness 0 to 4. The smaller the number, the brighter the star.

◁ *The Andromeda Galaxy is 2.2 million light-years away from us.*

Star map N I/1

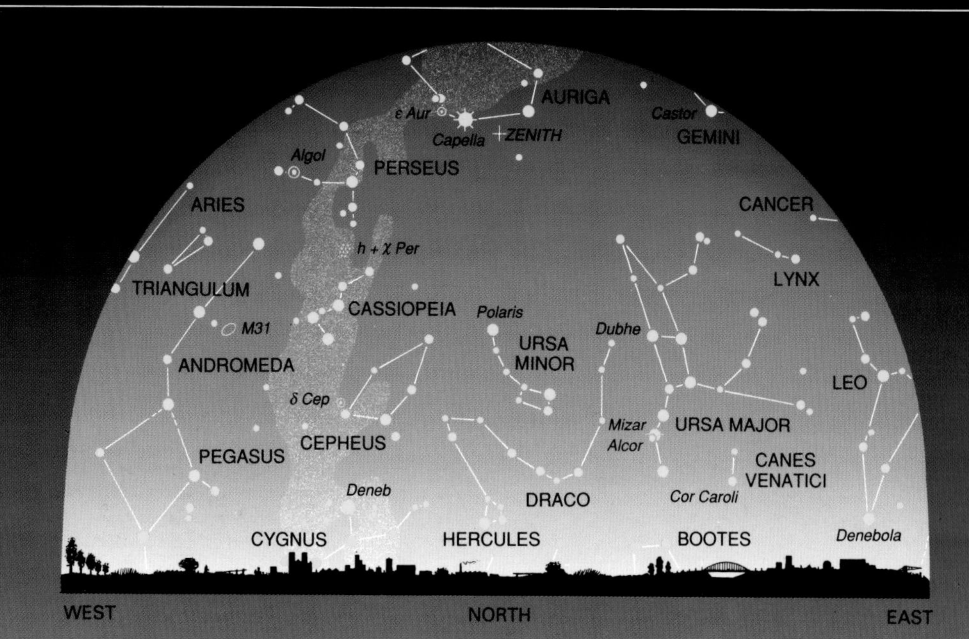

The evening skies of January offer some of the most beautiful views of the year. In the south, all the winter constellations are gathered. Particularly prominent are Orion, which includes an unusually large number of bright stars, and Sirius, the most brilliant star in the night sky. The Milky Way rises steeply out of the southeast and runs northwest, where, at the constellation Cygnus (the Swan), it dips again below the horizon. The ecliptic, which marks the apparent path of the Sun across the sky, and along which the planets, too, travel, is high in the southern sky, so that in many years the January sky is further adorned by bright planets.

The view to the north is equally interesting. In addition to Ursa Major in the northeast and the clearly visible Great Square of Pegasus, one other constellation stands out. Along with the Big Dipper it is one of the most widely recognized star patterns, namely **Cassiopeia**. Its five stars form an easily spotted W. This striking arrangement of stars gave rise to various imaginative interpretations from earliest times on. It has been taken to depict a hand, a bent leg, and—best known—a sitting female figure, which is in fact the basis of the constellation's present-day name.

In Greek mythology, Cassiopeia was a queen who, together with her husband, Cepheus, ruled over Ethiopia. They had a daughter, the beautiful Andromeda. Cassiopeia boasted that her daughter was fairer than the Nereids, the daughters of the sea god Poseidon. The Nereids complained of Cassiopeia's arrogance to their father, who sent a sea monster to devastate the coast of Ethiopia. In desperation King Cepheus sent a messenger to the oracle at Delphi for advice. The horrible answer that came back was that Ethiopia would get relief only if Andromeda was sacrificed to the sea monster. Not knowing what else to do, Cepheus and Cassiopeia followed the oracle's advice and chained Andromeda to a rock in the sea. The sea monster was al-

January 1, 11 P.M.
January 15, 10 P.M.
January 31, 9 P.M.

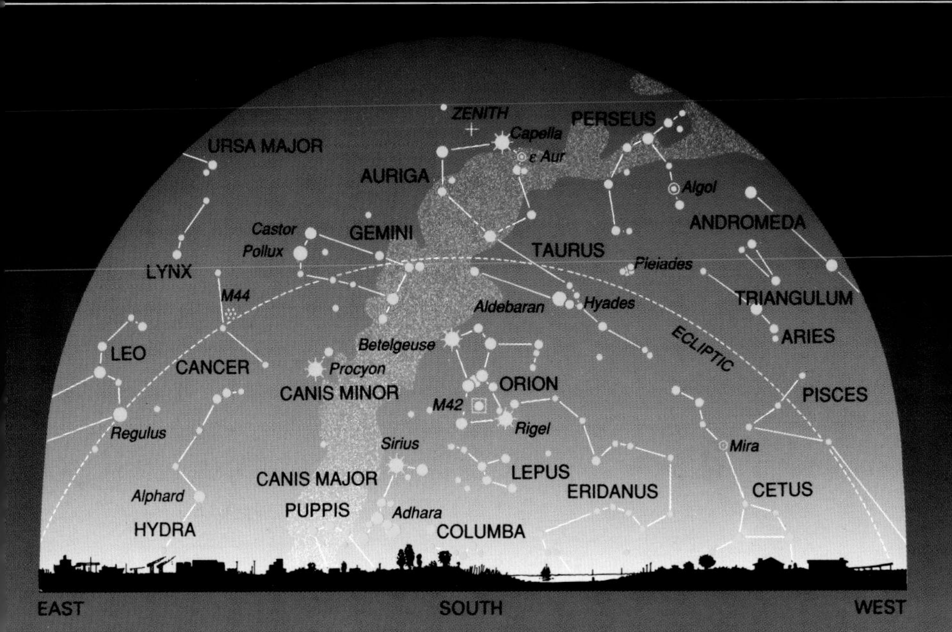

ready approaching and ready to devour Andromeda when the hero Perseus appeared on the scene. He challenged the monster to battle, and after a long, hard fight he subdued and killed it. He freed Andromeda and brought her back to her parents. Cepheus and Cassiopeia, not knowing how to thank him enough, gave him their daughter in marriage—Andromeda had fallen in love with her savior—and offered all of Ethiopia as a dowry. In commemoration of this great conflict, the gods placed all the participants in the heavens: Next to Cassiopeia and Cepheus we find not only Andromeda and Perseus but also the sea monster in the shape of Cetus, the Whale. The last is visible in January just above the southwestern horizon.

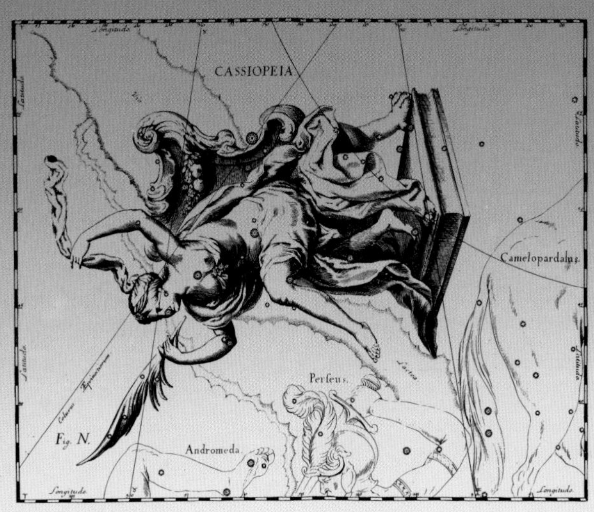

Cassiopeia

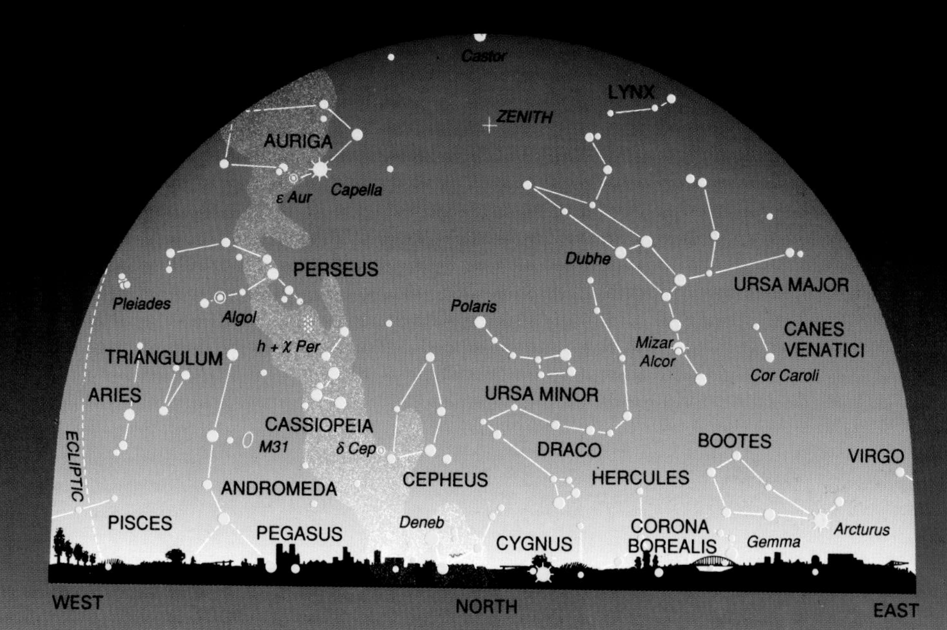

The northern sky is dominated by constellations associated with the stories from Greek mythology that tell about Cassiopeia, Cepheus, Andromeda, and **Perseus**. Perseus was the son of Zeus, the highest of the Greek gods, and Danaë. Because of his partially divine parentage Perseus was able to engage successfully in dangerous adventures, such as the rescuing of Andromeda from the clutches of the sea monster, which appears in the sky as the constellation Cetus (see pages 74–75). Another of his extraordinary deeds was the subduing of Medusa, a horror-inspiring creature, the sight of which turned anyone who beheld her to stone. The Greeks interpreted the constellation Perseus in such a way that the head of Medusa is exactly where the star Algol is, as depicted in the dramatic engraving reproduced from Hevelius's 1690 sky atlas. Algol, also called the demon star, fluctuates in brightness in a regular cycle (see page 130).

The constellations asso-

Perseus

February 1, 11 P.M.
February 15, 10 P.M.
February 28, 9 P.M.

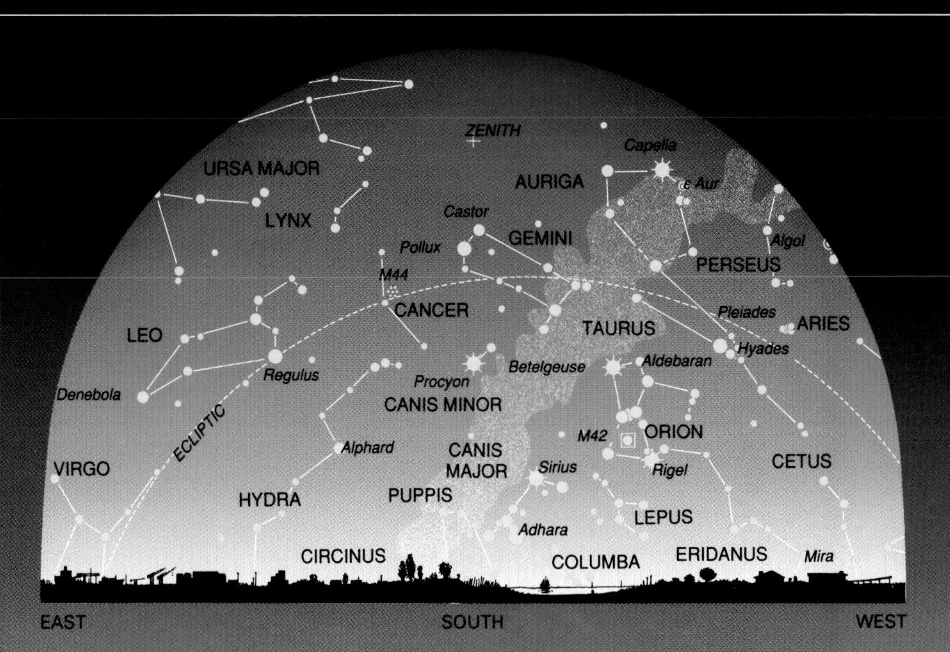

ciated with the stories about Perseus include a number of other conspicuous celestial objects that are easy to observe in February. The constellation Cepheus, for instance, includes the star Delta Cephei. It, too, is a variable star, although for reasons totally different from those that account for Algol's fluctuations in luminosity. Perseus contains the double star cluster h and χ (chi) Persei, which is a striking sight particularly when viewed through binoculars. Star clusters are collections of many hundred to several thousand stars that are bound together in space by mutual gravitational pull (see page 147).

The most spectacular and interesting celestial object is seen in Andromeda, namely, the great Andromeda Galaxy, M 31. M 31, which was in earlier times thought to be a nebula, is a gigantic galaxy removed from us by the unimaginable distance of 2.2 million light-years (see page 135). On clear nights one can distinctly make out a small, faintly luminous spindle shape (see photo on page 26). Every one of the light rays we perceive has been shooting through space for 2.2 million years before it reaches Earth. You can most easily locate M 31 by first finding the W of Cassiopeia (see January). Then, if you look closely, you can see that the right-hand trough of the W points like a small arrowhead toward M 31.

Over 400 years ago, in the year 1572, an exceptionally brilliant heavenly object turned up in the constellation Cassiopeia. It appeared to be a new star that shone brighter than Venus for several weeks, and it made an impression on the people of the time that is hard to imagine today. The Danish astronomer Tycho Brahe observed the new star especially carefully, which has gone down in history as Tycho's Star. Today we know that a supernova erupted in 1592 in that part of the sky, constituting one of the most gigantic celestial occurrences to have been observed to date (see Lexicon of Celestial Bodies, pages 149–50).

March

Star map N I/3

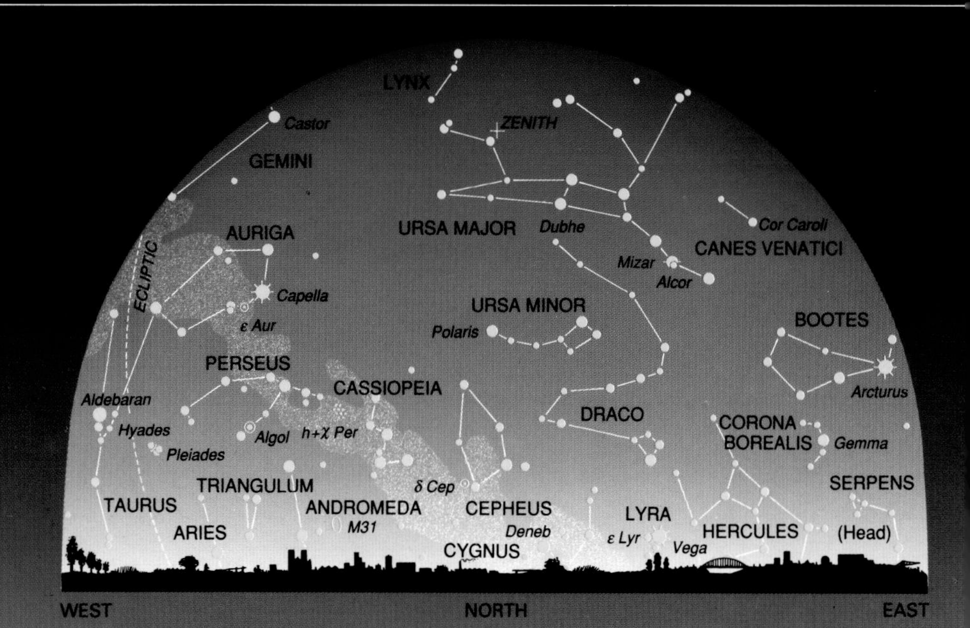

March ushers in spring in Earth's Northern Hemisphere. It is also the month when daylight saving time starts in many countries. The days now grow longer than the nights and, because the Sun sets later, the stars become visible later.

While the winter constellations, especially Orion, and the star Sirius sink below the horizon early in the evening in the southern sky, in the north **Ursa Major** (the Great Bear), including the seven prominent stars of the Big Dipper—probably the most generally recognized star pattern of the northern sky—assumes its highest and most conspicuous position. The Big Dipper is one of the few names that bear an obvious relation to the star patterns they refer to. There are four prominent stars that form the bowl or ladle of the Big Dipper; the three other stars are arranged in the slightly bent line and make up the handle. (In Europe the pattern is seen as a plough or a wagon—Charles's Wain.)

The official name of the entire constellation is Ursa Major, the Great Bear or, more accurately, the Great She-Bear. Many different legends are associated with this conspicuous constellation. The best known among them identifies the Great Bear as the nymph Callisto, a beautiful maiden with whom Zeus fell in love. Zeus's wife, the goddess Hera, was so angered that she turned Callisto into a she-bear and drove her away into the woods of Arcadia. She tormented Callisto so mercilessly that Zeus finally took pity on Callisto and transported her to the heavens as Ursa Major. He placed her far north in the sky so that she became a circumpolar constellation and he could behold her the entire year. What we see as the handle of the Big Dipper is the bear's tail, a rather long one for a bear, as the picture from Hevelius's atlas clearly shows, and quite unlike the short stubs earthly bears sport. In explanation of the elongated tail it is said that Zeus, eager to rescue his sweetheart from Hera's wrath, grabbed her by her nub of a tail and hurled her

March 1, 11 P.M.
March 15, 10 P.M.
March 31, 9 P.M.

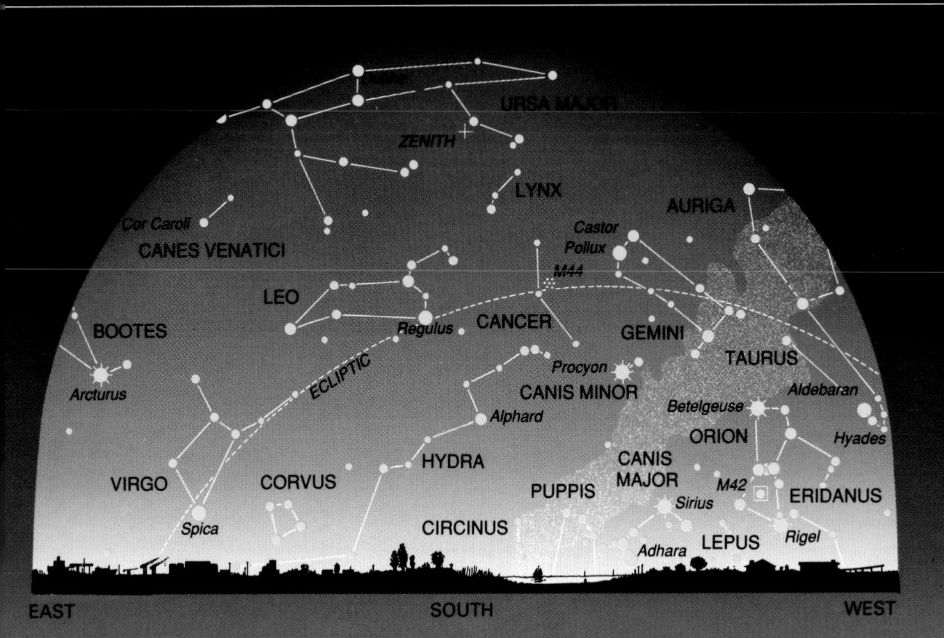

into the sky, pulling the tail and thus elongating it.

Ursa Major is of great help in locating other constellations because in the higher northern latitudes it is visible every night (being circumpolar) and because its major stars are easy to recognize when it is still dusk. Draw a mental line between the two stars forming the pouring end of the ladle of the Big Dipper and extend the line five times its length past the star Dubhe. This leads your eye to the North Star, or Polaris (see pages 22–23).

The middle star of the three forming the Dipper's handle (or the Bear's tail) is named Mizar and is especially well known. Right next to it one can make out a somewhat fainter star, Alcor, which was named **Eques Stellula**, or little horseman, in Latin. In this context the Dipper is pictured as a wagon or chariot with Alcor being a small rider perching on the chariot's shaft. Mizar and Alcor also are used to test one's visual acuity because it takes good eyes to see them as two separate stars.

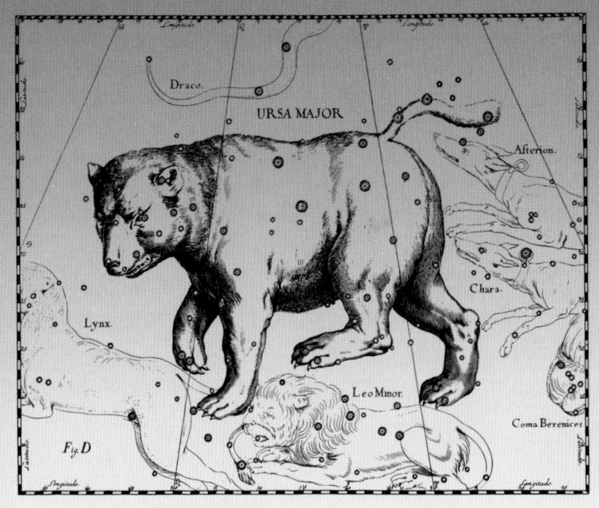

Ursa Major (Big Dipper)

April

Star map N I/4

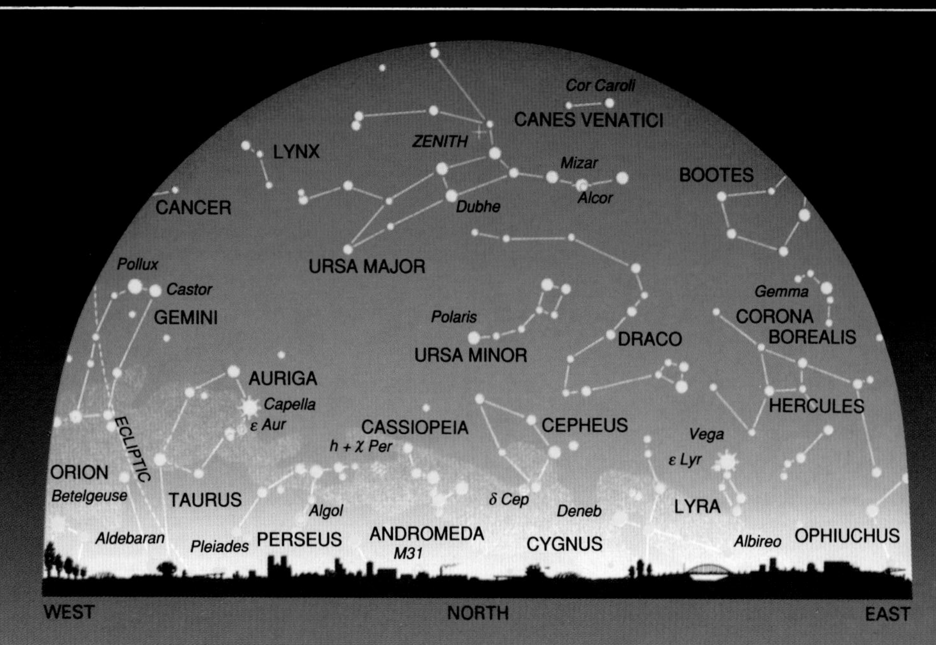

Starting in April, the northern constellations dominate the sky. In the south only three prominent stars are visible: Regulus, Spica, and Arcturus, together sometimes called the Vernal Triangle. Ursa Major and the constellations near it now reach their highest position in the sky, the Great Bear being, in fact, almost exactly at the zenith, directly above the observer's head.

With the help of Ursa Major it is easy to find one of the best known stars of the northern sky, Polaris. Polaris is the most prominent star of Ursa Minor, the Lesser Bear. It stands almost exactly at the north celestial pole and hardly moves at all in the course of the year (see pages 22–23). It is Polaris that Shakespeare refers to when he has Julius Caesar say to the potential murderers surrounding him:

But I am constant as the northern star,/Of whose true-fix'd and resting quality/ There is no fellow in the firmament./The skies are painted with unnumb'red sparks,/ They are all fire and every one doth shine;/But

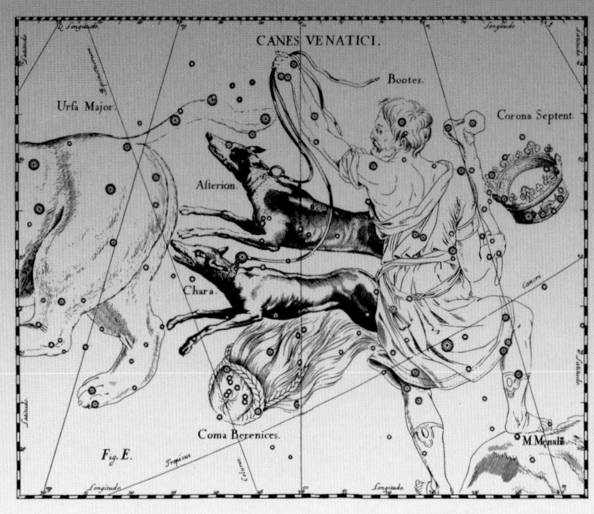

Canes Venatici

April 1, 11 P.M.
April 15, 10 P.M.
April 30, 9 P.M.

Add one hour if daylight saving time is in effect.

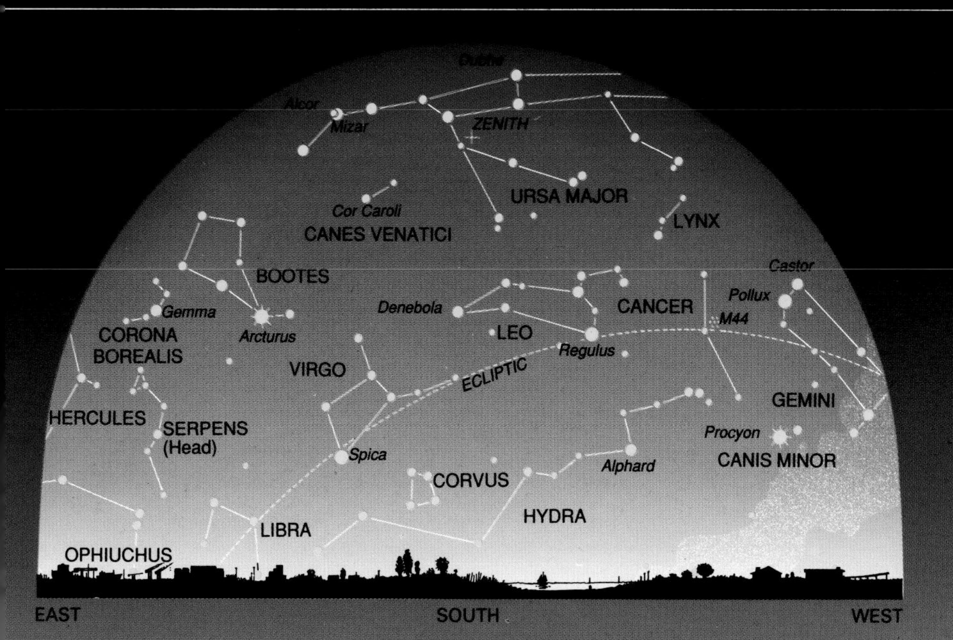

there's but one in all doth hold his place./According to modern astronomy, too, Polaris is unusual among the stars. It is a double star or binary, though its companion is visible only through a telescope. Its luminosity fluctuates slightly, and astronomers estimate that the light it emits—at a distance of about 360 light-years—is 1,600 times brighter than that of our Sun. The North Star is thus among the sky's giants.

Another constellation is visible in April high in the sky below the three stars forming the handle of the Big Dipper or the tail of the Great Bear. It is **Canes Venatici** (the Hunting Dogs). This is one of the few constellations seen in the Northern Hemisphere that does not go back to Greek mythology. It was first named around 1650 by Johannes Hevelius, from whose sky atlas the pictures of constellation figures in this book are taken. The brightest star in Canes Venatici is named Cor Caroli, which means Charles's heart. Most names of individual stars, such as Mizar, Alcor, Dubhe, and Arcturus, go back to Arabian astronomers of the first millenium A.D. Cor Caroli is an exception, having been named in 1725 by Edmund Halley, for whom the famous Halley's Comet is named (see pages 133–34). It is said that on May 29, 1660, when King Charles II arrived in London at the restoration of the monarchy after the brief Republican rule of Oliver Cromwell, this star shone especially brightly. However, this is nothing more than pretty popular legend. Cor Caroli has one of the strongest magnetic fields observed to date.

May

Star map N I/5

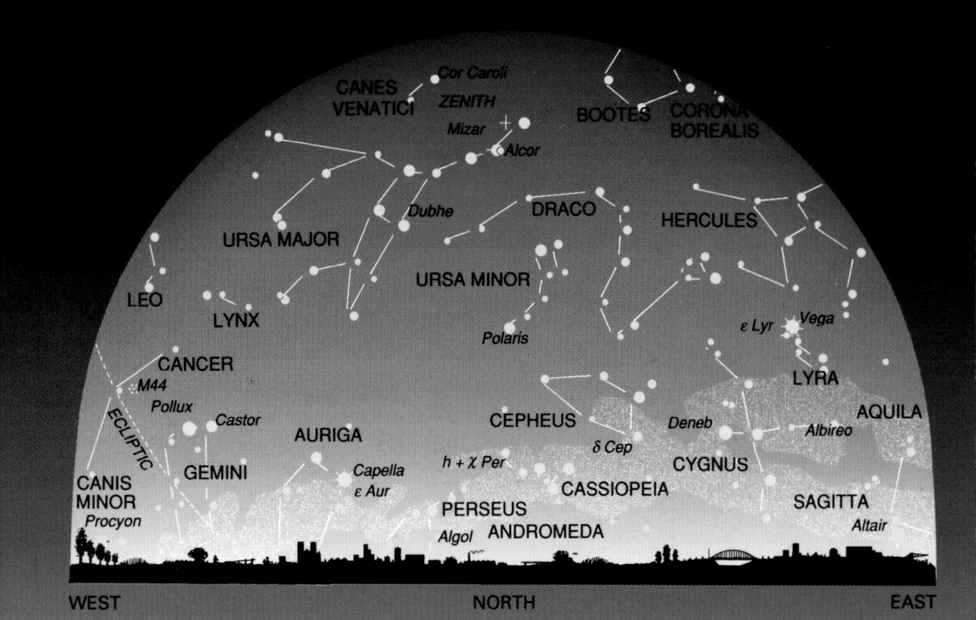

The night sky changes from month to month. Because the Sun moves through all the constellations of the zodiac in the course of the year (see the zodiac table on page 16), it constantly reveals different stars in a yearly cycle. As the constellations come and go, the position of the Milky Way, too, changes. Thus, in the northern sky, it lies low and parallel to the horizon in May. Visibility has to be especially good for it to be seen at this time of year.

In the southern sky the stars Arcturus, Spica, and Regulus shine forth, forming the Vernal Triangle. Arcturus is the brightest star in the northern half of the sky. It is one of the first stars that can be detected shortly after dusk, and, with the help of the Big Dipper, it is easy to find. If the arc of the Dipper's handle, formed by three stars with Mizar in the middle, is extended across the sky, it leads inevitably to Arcturus, and if the arc is extended even farther, the eye is led to Spica, too.

Arcturus is part of **Boötes** (the Herdsman), a constellation with which several different myths are associated. According to one, Boötes is driving Canes Venatici (the Hunting Dogs), across the sky, chasing them after the Great Bear and the Lesser Bear. Another, prettier story links Boötes to the constellation Virgo (see pages 38–39).

Next to Boötes is the Northern Crown, or Corona Borealis (in the sky of the Southern Hemisphere there is a Southern Crown). This is one of the few instances where no great imagination is required to explain the constellation's name. The Northern Crown's five stars are arranged in a semicircle with one that sparkles especially brightly—named Gemma—in the middle. Gemma, which means "jewel," is 75 light-years distant. It is a so-called spectroscopic binary, that is, a double star whose two components cannot be seen separately even through the most powerful telescope. Only an analysis of its spectrum reveals the star's double nature. The two component

May 1, 11 P.M.
May 15, 10 P.M.
May 31, 9 P.M.

Add one hour if daylight saving time is in effect.

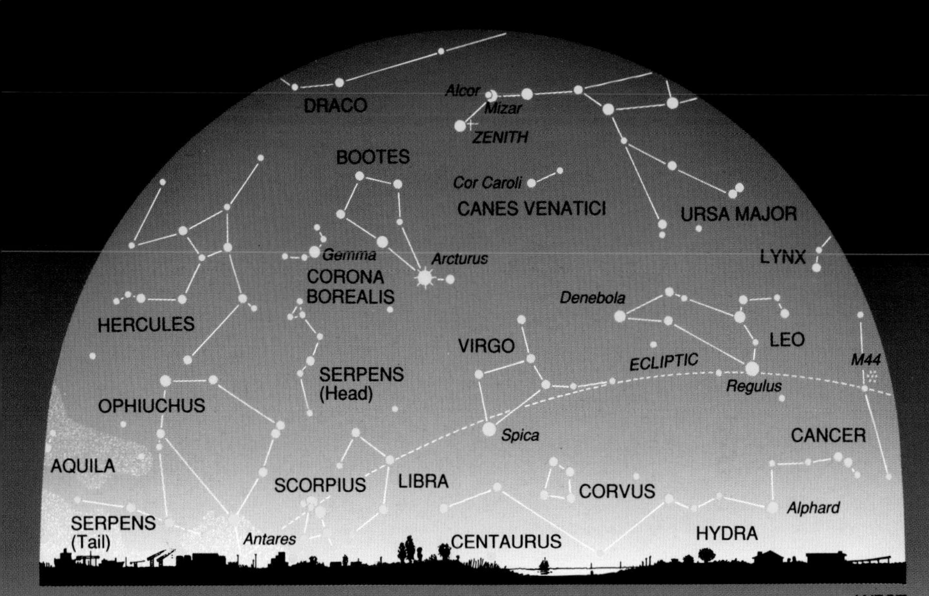

stars of Gemma revolve around each other once in 17 days and 12 hours.

The Northern Crown is associated with the myths surrounding Theseus. Theseus accomplished his greatest deed on the island of Crete, where he destroyed the monstrous Minotaur. The Minotaur lived in a labyrinth in the capital city of Crete, and every nine years seven youths and seven virgins had to be sacrificed to it. Theseus decided to kill the Minotaur. Ariadne, the king's daughter, helped him by giving him the famous ball of string. Theseus tied one end of the string to the entrance of the labyrinth and unrolled the ball as he went. Thus, after killing the monster, he was able to find his way out of the labyrinth by following the string. While embarking on this dangerous venture, he wore a crown that the gods had given him and from which shone a brilliant jewel. After slaying the monster, Theseus gave the crown to Ariadne, and later the gods placed it, including Gemma, in the heavens.

Boötes

June

Star map N I/6

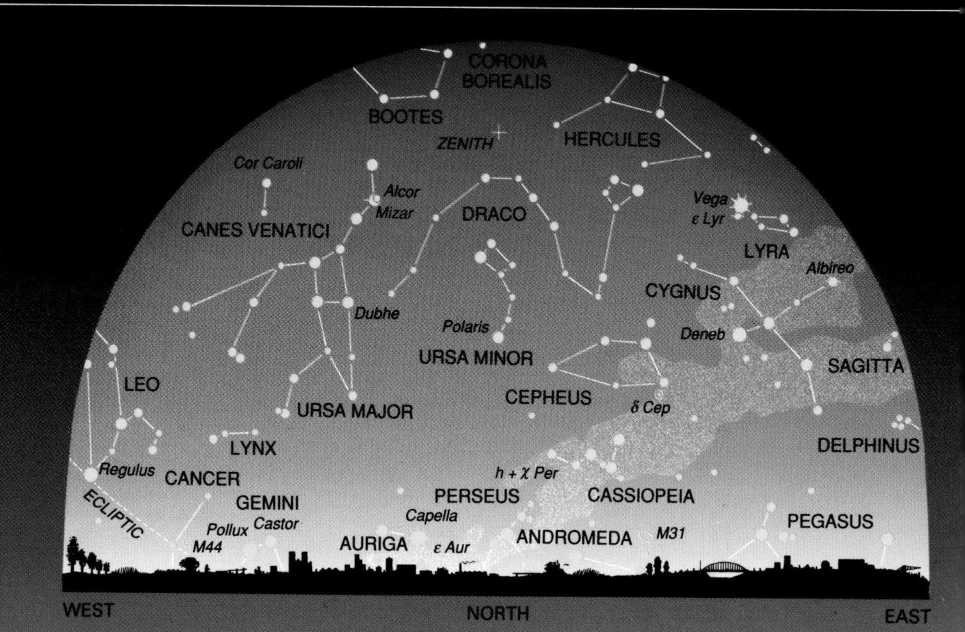

Starting in June, the visibility of the stars is seriously affected by the onset of summer. The Sun sets so late and twilight persists so long (see Dusk, Midnight Sun, and Polar Night on page 12), that the night is dark enough for watching stars for only about five hours or less.

In the sky, **Virgo** (the Virgin) and Boötes are next to each other, and their stories are bound together in myth, too. In addition to the story claiming that Boötes is herding the bears with his hunting dogs, there is another one. In it, Boötes, the Herdsman, is identified with a mortal named Icarius to whom the gods had presented the gift of wine. Icarius traveled through the countryside with a cart, offering the new beverage to the public. But because they were not familiar with wine, people thought Icarius was trying to poison them and killed him. Only later did they realize what a delightful drink wine was, and then they deeply regretted what they had done. Soon thereafter Icarius's daughter came along looking for her father. When she learned about the sad fate that had befallen him she wept bitter tears, and the gods took pity on both the slain father and his virgin daughter, transforming them into constellations. Even the cart with which the herdsman Icarius had traveled found a place in the heavens as the constellation more generally known as Ursa Minor or the Lesser Bear, which is in this context referred to as the Lesser Wagon. Because Icarius's daughter had carried some ears of wheat in her hand while searching for her father, the brightest star of Virgo is called Spica, which means ear of grain.

As this story about Icarius the Herdsman shows, the constellation now called Ursa Minor was known already among the Greeks as the Lesser Wagon, and, because of its similarity to the Greater Bear, or Greater Wagon, the name Lesser Wagon is sometimes still popularly applied to it in Europe. This group of stars is also the famous Little Dipper. Virgo, with the star Spica, is one of the zodiacal constellations, and the eclip-

June 1, 11 P.M.
June 15, 10 P.M.
June 30, 9 P.M.

Add one hour if daylight saving time is in effect.

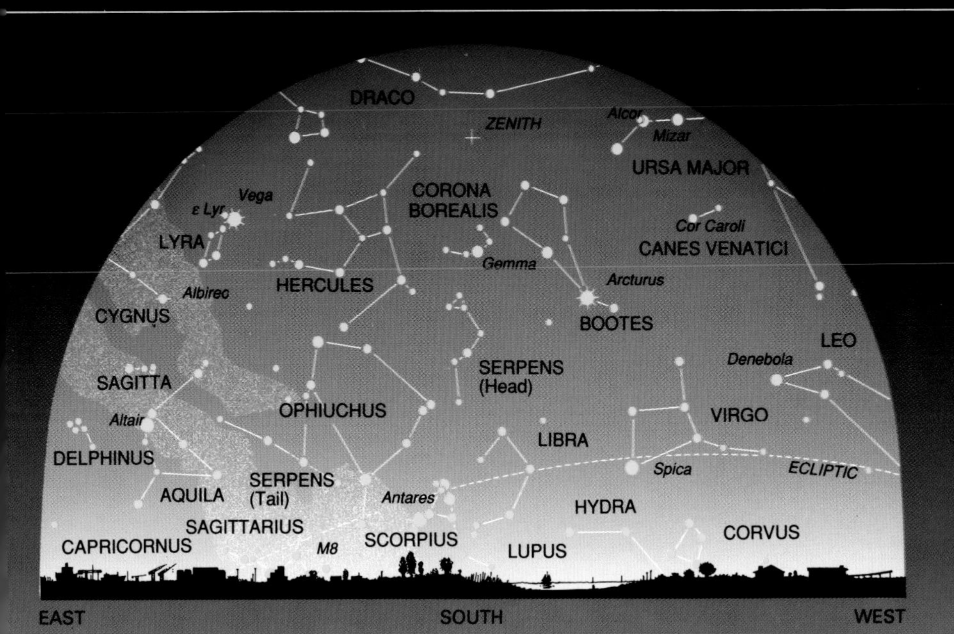

tic cuts across it. The Moon and all the planets are seen only near the ecliptic and in these constellations (see The Moon and the Planets on page 16).

Spica is an exceptionally radiant star, and it is only because of its great distance from Earth that it is not counted among the brightest stars of the sky. Spica is 300 light-years distant, and its intrinsic luminosity is about 2,300 times greater than the Sun's. It is a so-called spectrographic binary (see page 132), which means that it is really made up of two stars that circle around each other, in this case taking four days for one complete revolution. They are separated from each other by 13 million miles (20 million km), a distance that sounds immense but that shrinks into an angle too small to be measured because the stars are so far removed from us.

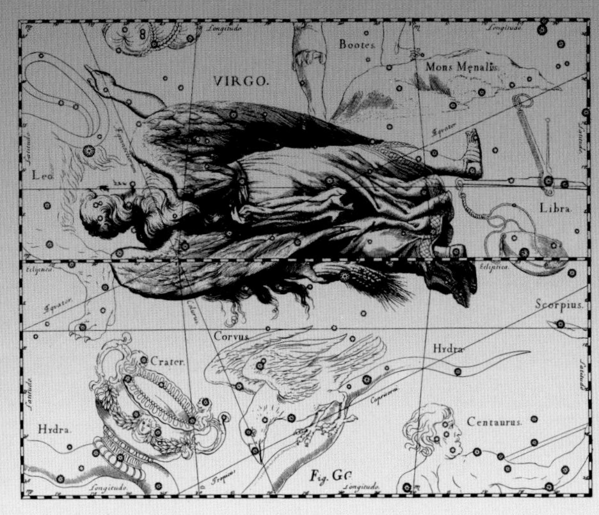

Virgo

July

Star map N I/7

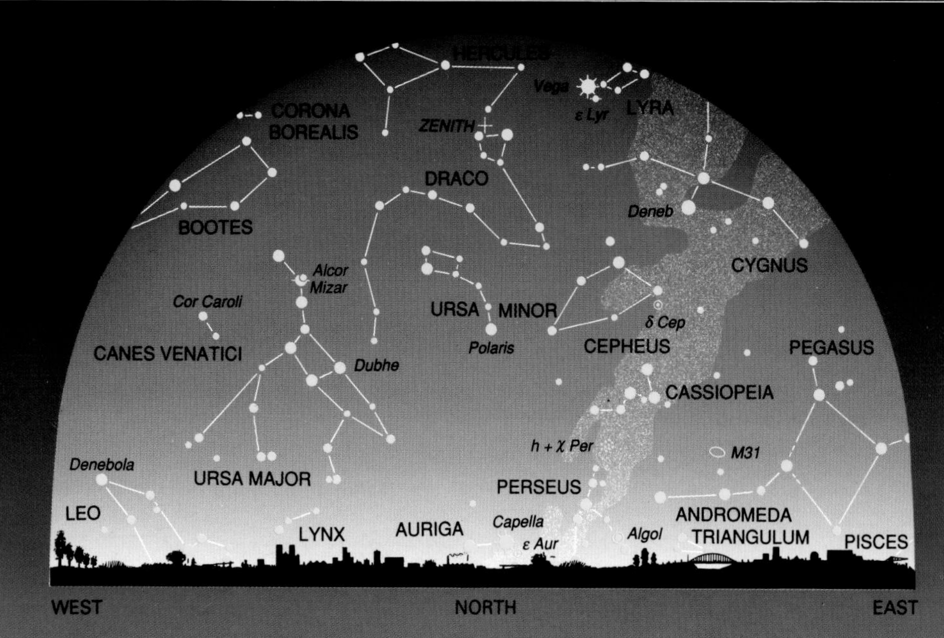

In July Earth is at its greatest distance from the Sun (see page 13). The nights are short, and the ecliptic lies low in the south. Consequently the Moon and planets, too, don't rise far above the southern horizon during July. This is why the full moon is always low in the southern sky during summer, whereas in the winter, when the ecliptic is high in the sky, the full moon traverses the firmament more prominently and higher up.

While the ecliptic and, consequently, the constellations of the zodiac are close to the horizon, the Milky Way is all the higher overhead in the summer night sky. The summer months are the best time to observe the Milky Way in the Northern Hemisphere. Today we know that the Milky Way is made up of the combined light of millions of stars, stars that are too faint to be seen individually (see page 140). Before the invention of telescopes, the Milky Way seemed almost like a miraculous object, although many astronomers and thinkers of ancient times suspected its true nature.

There is an old story about the origin and name of the Milky Way that is connected with the famed hero of Greek mythology, **Hercules**. His constellation is in its best viewing position in the sky during the evening hours of July. Hercules was the son of Zeus, but his mother was not Zeus's wife, Hera, but instead the fair mortal Alcmene, daughter of the Mycenean king. Wanting to endow his son with divine strength, Zeus placed the newborn against his sleeping wife's breast, where he could imbibe divine mother's milk. Hercules, however, was so strong even as a baby and sucked so vigorously that the milk spurted way up into the air, rising from Mount Olympus, the abode of the gods, across the sky. The trail it left became the Milky Way. Later on, he performed the 12 Herculean labors as well as other heroic deeds, earning him immortal fame and raising some of his opponents to the status of heavenly constellations as well.

July 1, 11 P.M.
July 15, 10 P.M.
July 31, 9 P.M.

Add one hour if daylight saving time is in effect.

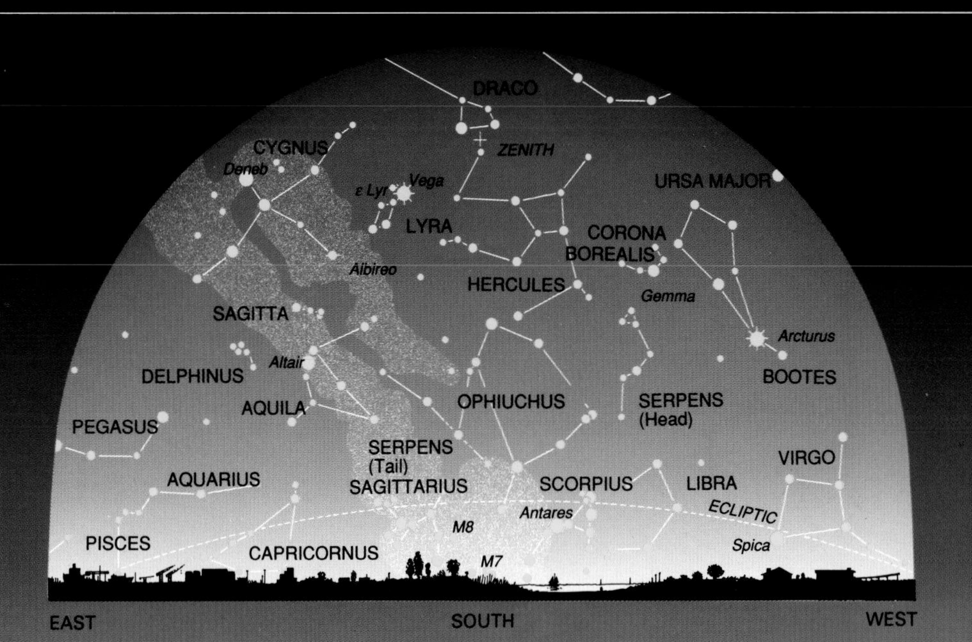

The constellation Hercules lies east of Boötes and the Northern Crown (Corona Borealis) and is easiest to find if one looks for the trapezoid formed by the brightest stars at the center of the constellation. To the east of Hercules, in the constellation Cygnus (the Swan), the Milky Way—created according to myth by Hercules—shows a strange feature. It splits into two bands with a dark, starless area between them. This apparent fork in the Milky Way is caused by dark dust clouds that intrude between us and the millions of stars behind them, blocking the light of these stars. The dark clouds of the Milky Way show up dramatically in photographs of the Milky Way and are particularly conspicuous in the constellation Cygnus (see photo, page 140).

Hercules

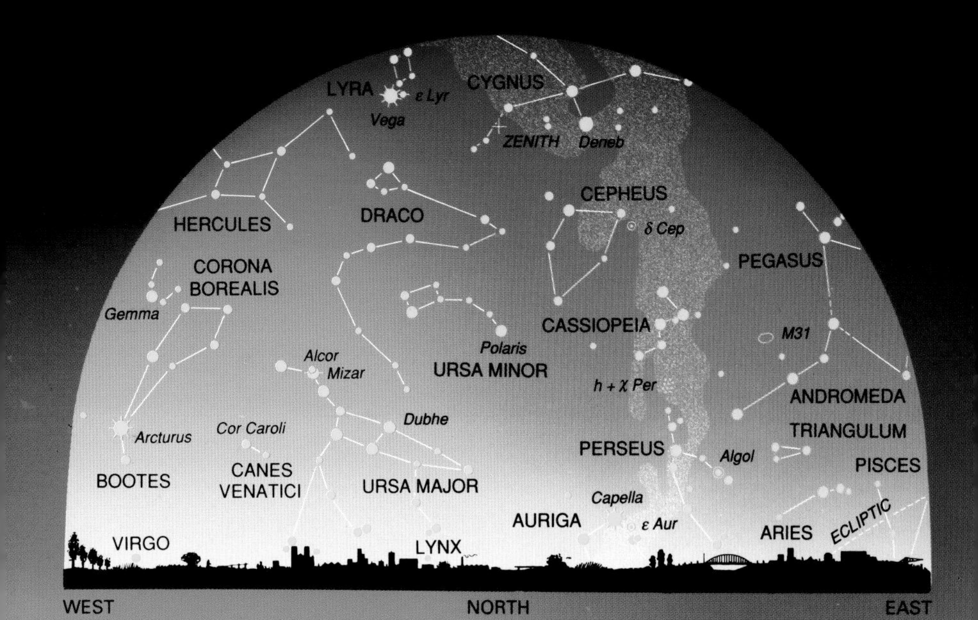

Hercules had to perform 12 labors before he was to be permitted to become king of Argos. The victim of his twelfth and last labor is seen in August in a favorable position in the northern sky high above the horizon: It is the constellation **Draco** (the Dragon).

According to Greek myth, a dragon guarded the golden apples of the Hesperides. No one had been strong enough to overcome this dragon until Hercules, after long, errant traveling, arrived at the tree planted by Hera and slew the beast with his gigantic club. He took the apples and brought them back to Eurystheus as proof that he had successfully accomplished the last of his labors. Hera gave the dragon a place in the heavens as a reward for having guarded her tree.

The place for this beast could not have been chosen better, for the stars forming the dragon's body coil through a starless region near the north celestial pole, and the head is marked by four stars next to Hercules. Old representations of the sky's constellations, such as the one in Hevelius's atlas, show the foot of Hercules resting on the conquered beast's head. Hevelius's picture illustrates two other interesting aspects of the constellation. The north pole of the ecliptic is located in Draco. The ecliptic—the apparent path the Sun takes across the sky (see Everything about the Apparent Path of the Sun, page 15)—has a north and a south pole. This is analagous to the celestial equator and Earth's equator, which have a north and south pole on an axis exactly perpendicular to their plane. In Hevelius's picture, a circle is drawn around the ecliptic north pole in a black-and-white band that runs through the celestial north pole. The distance between the north celestial pole and the ecliptic pole corresponds exactly to the angle by which Earth's axis deviates from the perpendicular of the plane of its orbit around the Sun. The north celestial pole moves around the ecliptic pole describing the circle shown in Hevelius's picture in approximately 26,000 years.

August 1, 11 P.M.
August 15, 10 P.M.
August 31, 9 P.M.

Add one hour if daylight saving time is in effect.

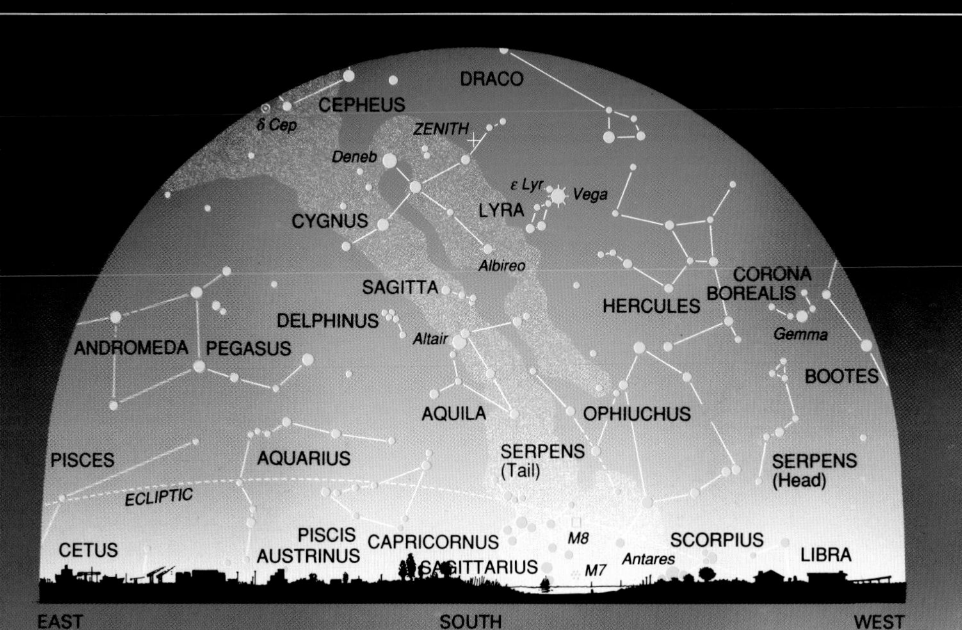

Astronomers refer to this important phenomenon as precession, that is, a moving ahead of the celestial pole. This amazing situation, which was recognized in early times by Greek astronomers, accounts for the fact that the north celestial pole slowly moves from constellation to constellation. At the moment it is located at the north or pole star in the constellation Ursa Minor. About 4,000 years ago it lay within Draco, close to a star below the two brightest stars in the Lesser Bear's body (that is, the two outer stars of the bowl of the Little Dipper). In Hevelius's chart, this star is drawn quite prominently. Its name is Thuban. About 5,000 years ago it was the pole star, although it was a rather feeble beacon because it is considerably less bright than Polaris, our present-day North Star.

Draco

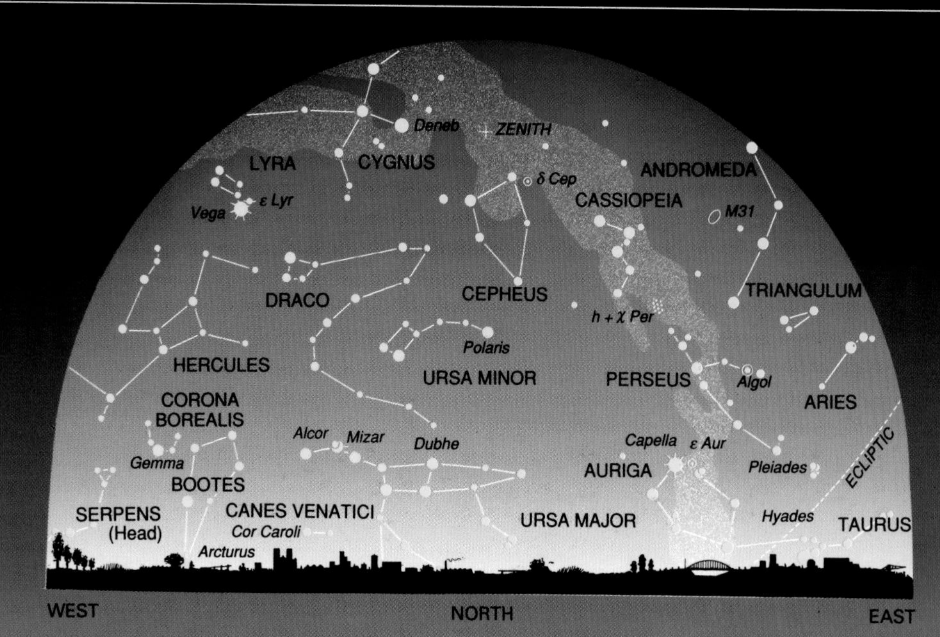

In the Northern Hemisphere, summer officially comes to an end during this month (see page 104). The nights again become longer than the days, and visibility is noticeably improved even during the early evening.

The summer sky is marked for the viewer in the Northern Hemisphere by the Summer Triangle, which is not a constellation but simply a configuration useful for getting oriented. The triangle is made up of Deneb, Altair, and Vega, the brightest stars in the constellations Cygnus, Aquila (the Eagle), and Lyra, respectively. The Summer Triangle is especially helpful because its stars can be perceived quite clearly even in the lit up skies over big cities.

Lyra (the Lyre) and Cygnus (the Swan) are interesting primarily on account of the double stars they contain. Epsilon Lyrae is a double star. It and Mizar-Alcor in Ursa Major (see pages 32–33) are the only binaries whose two component stars are discernable by the unaided eye. Epsilon Lyrae in particular is a real test of eyesight because only someone with extremely sharp vision is able to make out that it is a double star. If you reach for the binoculars, you can tell immediately that Epsilon Lyrae is made up of two stars that are quite far apart from each other. The two stars, also called Epsilon 1 and Epsilon 2, are in turn binaries, but this can be seen only with the aid of a telescope (see page 133).

Cygnus is sometimes referred to as the northern cross because its stars line up in the shape of a cross. It contains the second double star mentioned above, Albireo, but in order to see its two stars separately you need binoculars.

Vega in the constellation Lyra is one of the brightest stars in the northern sky, being almost as brilliant as Arcturus (see page 131). Deneb in the constellation Cygnus looks fainter than Vega, but this is only because Vega is much closer to Earth than Deneb. Vega is a star of normal size, 58 times brighter than the Sun and 26 light-years distant (for more on Vega, see page 151). Deneb,

September 1, 11 P.M.
September 15, 10 P.M.
September 30, 9 P.M.

Add one hour if daylight saving time is in effect.

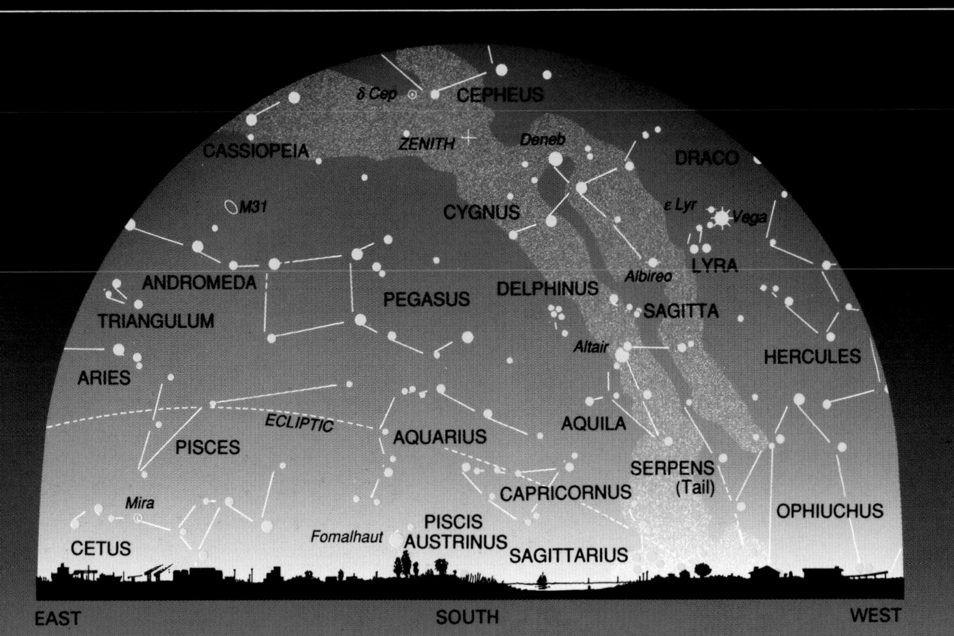

by contrast, can be described only in superlatives. Astronomers call big stars like Deneb supergiants, and Deneb is, in fact, of gigantic dimensions. Deneb's radiance is at least 60,000 times that of the Sun, and it is 1,600 light-years removed from us, the greatest distance of all the bright stars of the first magnitude. If Deneb were the same distance away from us as Vega is, it would be the brightest star in the sky—16 times brighter even than Venus, presently the most luminous celestial body. It is only the huge distance of 1,600 light-years that makes Deneb's overpowering light output shrink to a moderate brightness that is comparable to that of stars located much closer to us. The only reason why all the stars in the sky appear to us equidistant is that our senses and mind are incapable of perceiving or comprehending the astronomic distances involved.

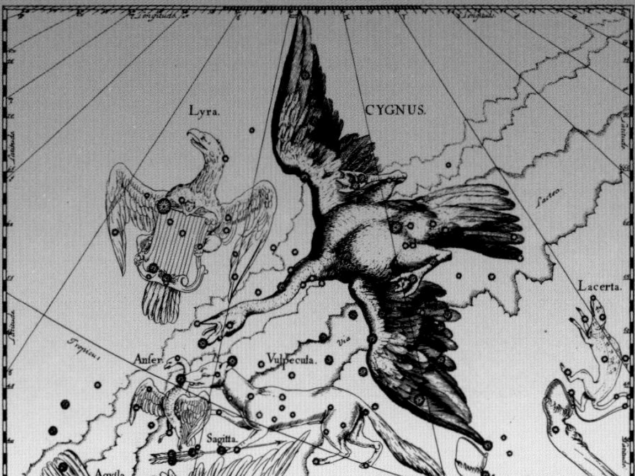

Cygnus

October

Star map N I/10

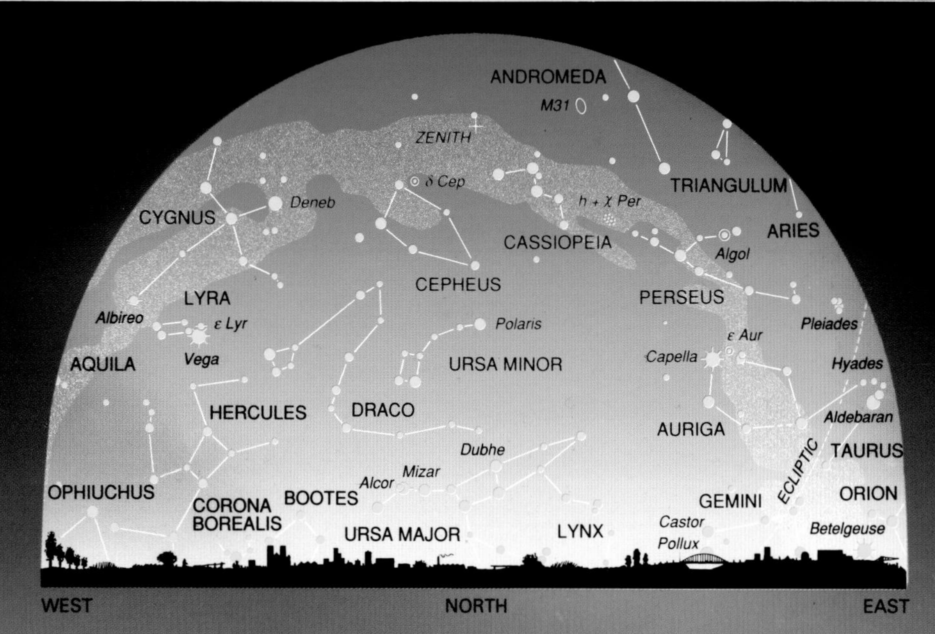

In October, the Summer Triangle can be seen in the west during the early evening hours. The constellations Lyra, Cygnus, and Aquila, which contain the stars making up the Summer Triangle, also go back to Greek mythology. The lyre was the instrument played so beautifully by the famous singer Orpheus that even the stones were moved to weep. Cygnus is usually associated with the story of Leda and the swan. Leda was a beautiful mortal with whom the god Zeus fell in love. Because Leda loved her husband and was true to him, Zeus approached her in the guise of a swan. Leda, who did not recognize Zeus, entered the water with him, and from their union beautiful Helen was born, on whose account the Trojan War was fought. In 1975 a new star or nova rose to brightness in Cygnus and could be seen easily by the unaided eye for several weeks, even making headlines in the newspapers. **Aquila** (the Eagle) was sent by Zeus to sweep up the youth Antinous and bring him up

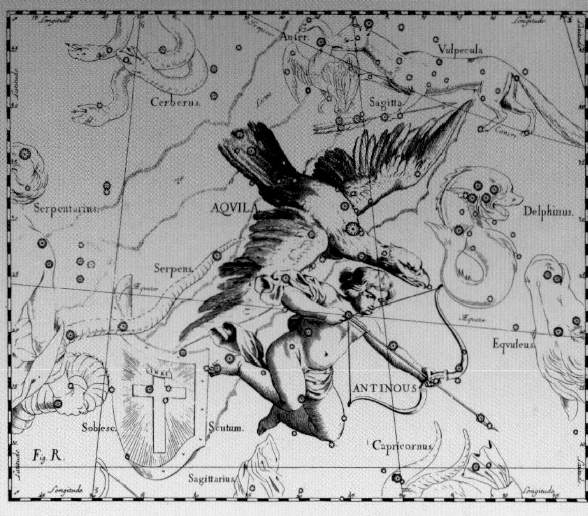

Aquila

October 15, 10 P.M.
October 31, 9 P.M.

Add one hour if daylight saving time is in effect.

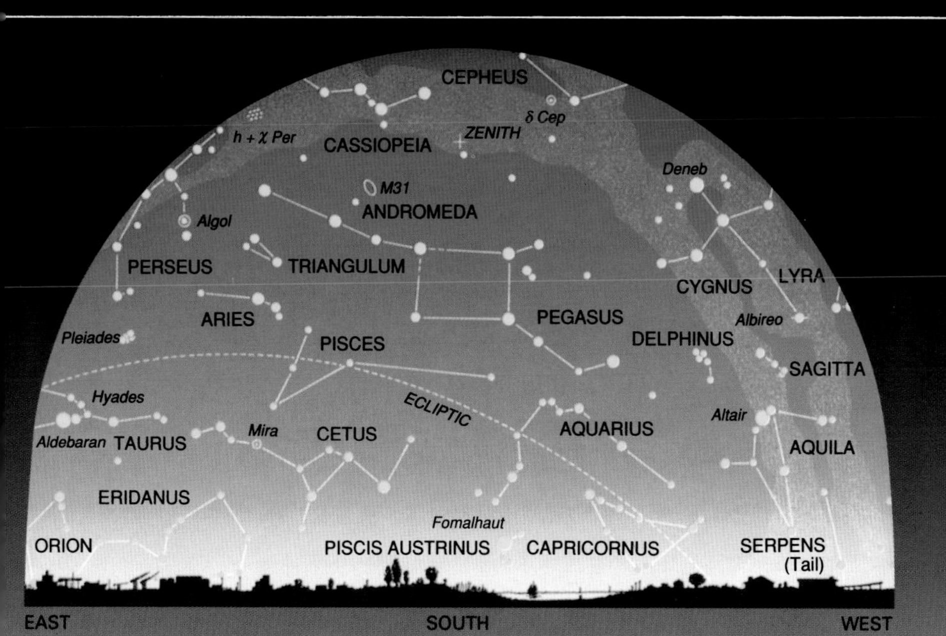

to Olympus, where he became cupbearer for the gods. As an expression of gratitude the gods originally placed both the eagle and Antinous among the stars, and in Hevelius's atlas both figures appear together. But as a consequence of an overall revision of the constellations that became effective in 1930, Antinous has vanished from today's sky.

South of Cygnus there are two small constellations. One of them, Delphinus (the Dolphin), is very easy to spot. Its four brightest stars form an easily recognizable trapezoid. The other one is Sagitta (the Arrow), one of the smallest of all the constellations. The dolphin, according to legend, helped Poseidon find lovely Amphitrite, whom the sea god had long been looking for in vain. In his gratitude, Poseidon transformed the dolphin into a heavenly constellation. The arrow, finally, is said to have been used by Hercules to kill an eagle that was tormenting Orpheus. (This is obviously a different story from the one in which the eagle is connected with the boy Antinous.)

The brightest star in Aquila (the Eagle) is Altair. It takes only 6½ hours to spin once around its axis (our Sun takes 25½ days; see pages 149–50). Altair is 16 light-years away from Earth (that amounts to roughly 100 trillion miles or 160 trillion km) and is thus one of our Sun's closer neighbors. Much more spectacular than Altair was Nova Aquilae, a nova that burst forth in great brightness in June of 1918, lending Aquila a very different look for several weeks as it grew more luminous than Sirius, the brightest star in the sky. Novas (new stars) are sudden outbursts of light in previously obscure stars. These dramatic surges are due to profound changes in the physical processes going on in the stars (see pages 142–43).

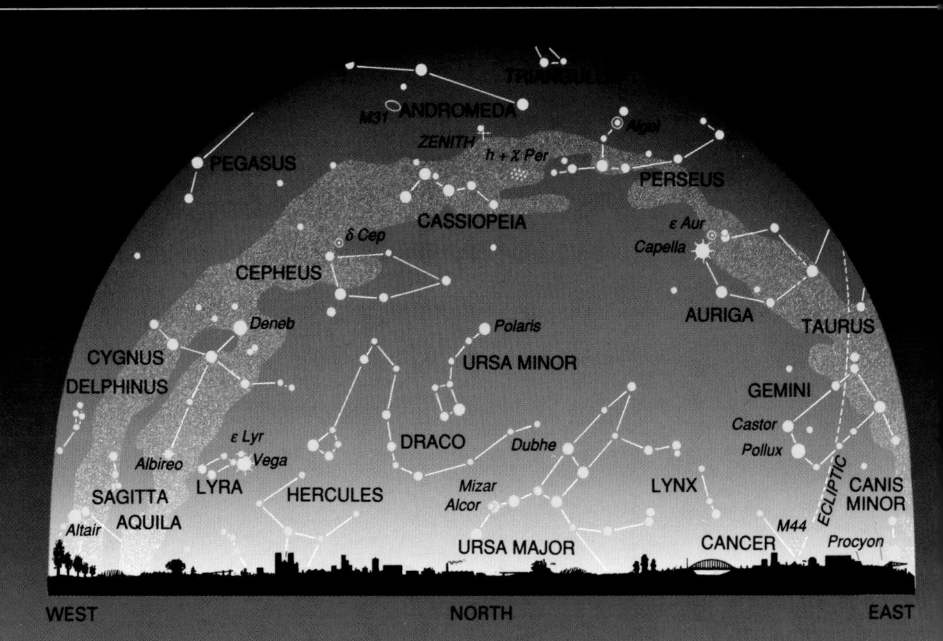

The approach of winter becomes apparent during November not only in the air but in the sky as well. In the evening, Orion, the constellation typical of the winter sky, already can be seen in the east. Above Orion we find, clearly visible, **Auriga** (the Charioteer) and Taurus (the Bull), two more winter constellations. To be sure, Auriga has been quite apparent in the sky for some months; in fact, Capella, its brightest star, is circumpolar and thus can be seen throughout the year (see page 23).

Capella is one of the ten brightest stars in the sky (see page 133). Capella, which means little she-goat, usually is depicted in old renditions of the starry sky as sitting on the shoulder of the charioteer, who holds a whip. At this point in time it is difficult to understand just what the connection between the charioteer and the little goat originally was. A number of sometimes mutually contradictory stories surround the figure of the charioteer.

According to one of these stories, the charioteer was Phaeton, the son of the sun god Helios. Helios drove across the sky every day in a fiery chariot, providing light for Earth. But only he could control the powerful horses that pulled the chariot. When Phaeton, going against his father's express prohibition, tried to drive the chariot, he lost control of the horses and tumbled from the sky. The course the plunging, careening chariot took is said to have formed the Milky Way—a totally different explanation from that offered by the legend that tells of milk spurting across the sky from Hera's breast (see page 40).

In another myth Auriga, the charioteer, represents Hephaestus, the god of fire and iron forging. He invented the four-horse chariot and, as a reward, was transformed into a constellation.

Auriga appears in the sky as a figure made up by four stars. Usually a fifth star to the south is included even though the exact delineation of the constellations reflecting modern revisions assigns it to Taurus. That is why this

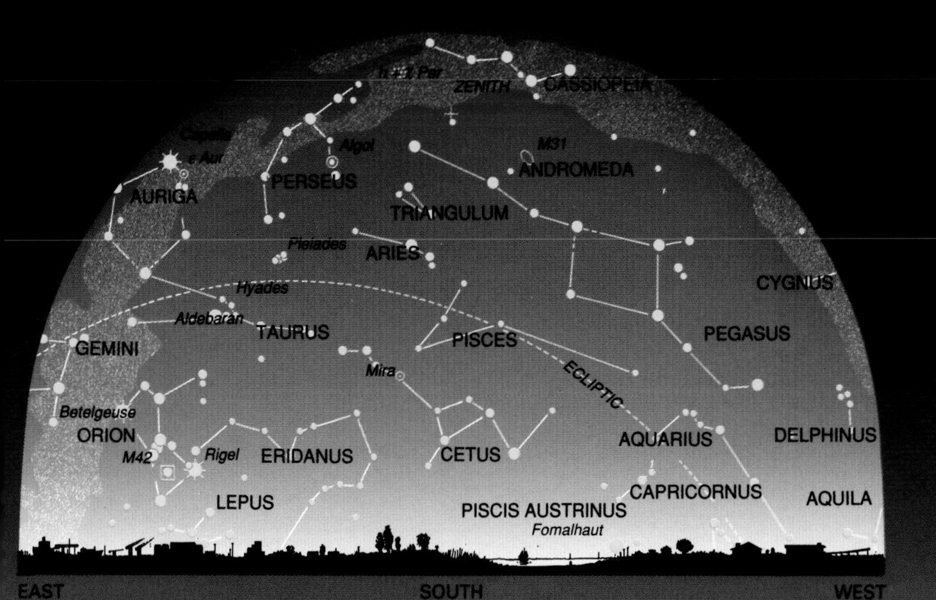

star is connected to the rest of Auriga with dotted lines. In addition to bright Capella, Auriga contains another remarkable star, Epsilon Aurigae. It is a variable star of huge dimensions, one of the largest and most mysterious variables known (see pages 150–51).

Every November a shower of meteors is seen, the so-called Leonid meteors. They appear in the middle of the month, and approximately every 33 years there is an unusually large number of them (see page 139). The Leonids all seem to radiate from the constellation Leo (the Lion). Leo does not appear in November's sky charts because it does not become visible until early morning, around 2 A.M., and different charts apply for that time (see table, page 20).

Auriga

December

Star map N I/12

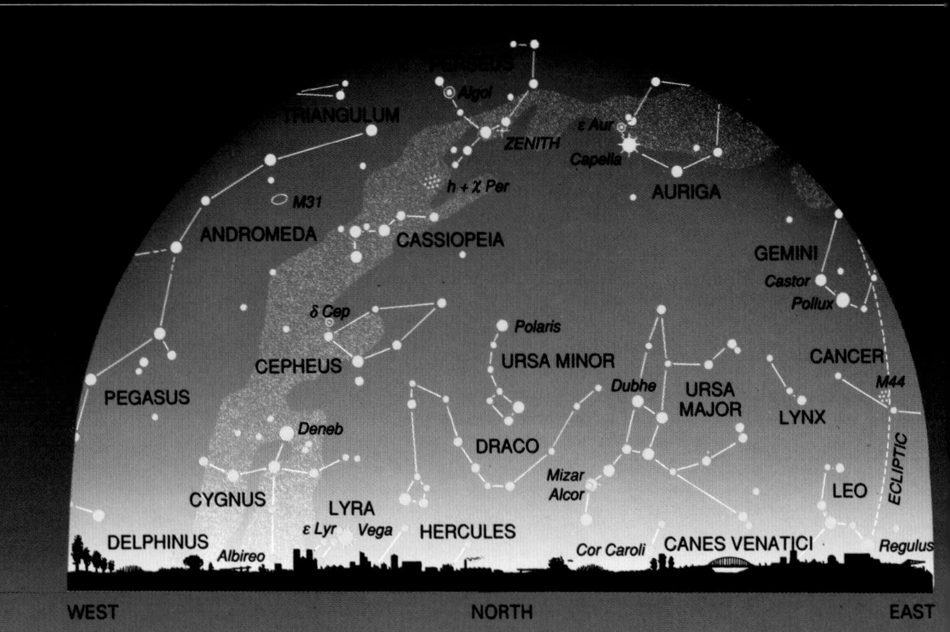

December marks the beginning of winter in the Northern Hemisphere. In the evening all the winter constellations are gathered in the east, and their brightest stars combine in a large hexagon, the Winter Hexagon. The corners are made up by the stars Rigel in Orion, Sirius in Canis Major, Procyon in Canis Minor, Castor and Pollux in Gemini, Capella in Auriga, and Aldebaran in Taurus.

Aldebaran belongs to the groups of red giants (distance: 68 light-years; diameter: 40 times that of the Sun) and appears reddish in the midst of the Hyades star cluster, which it is not, however, part of. Farther north is the Pleiades star cluster. It is part of the constellation **Taurus**. The Hyades and the Pleiades are the most conspicuous and famous of the star clusters (see pages 135, 145, and 147).

Aldebaran is one of the major reasons why the constellation Taurus was conceived, for not only the Greeks but many other early people thought they could see a powerful beast in this group of stars, with reddish Aldebaran forming its eye. Aldebaran, along with Spica in the constellation Virgo (see pages 38–39), is one of the few bright stars that can be eclipsed by the Moon. This is possible because Aldebaran stands immediately next to the ecliptic, the apparent path the Sun follows across the sky.

The most famous of the eclipses of Aldebaran took place in the year A.D. 509. It was visible from Athens and gave rise, 1,200 years later, in the early days of modern astronomy, to an important discovery. In 1735, Edmund Halley reconstructed the course of the Moon far back into the past. He found to his amazement that according to his calculations Aldebaran could not have been eclipsed by the Moon in A.D. 509—Aldebaran would have had to be farther to the south. This inspired Halley's idea—a true stroke of genius considering the time when he conceived it—that Aldebaran might have changed position in the course of almost 1,200 years. Halley was thus the first to discover that the

December 1, 11 P.M.
December 15, 10 P.M.
December 31, 9 P.M.

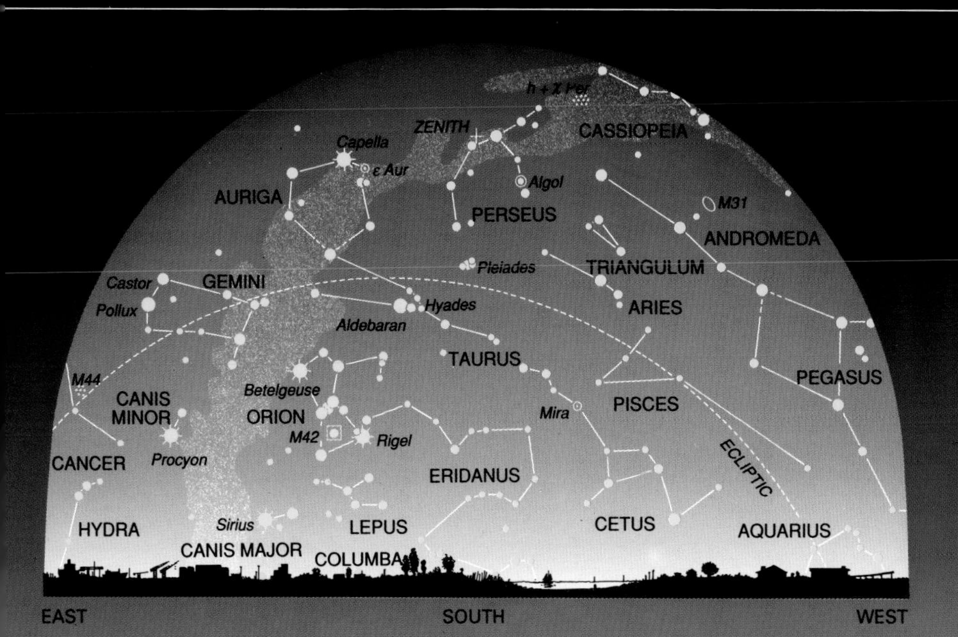

so-called fixed stars are not stationary. In fact, the stars only seem to adhere to their places in the sky. We get that impression because they are so astronomically far away from us. The stars' remoteness makes the vast distances they travel shrink into angles that are too small to measure. All stars are in motion, but their movement is discernable only by means of precise measurements or through observation over a very long time span—over 1,000 years in Halley's case. The movement of the stars has produced no visible changes in the sky since the Greeks scanned it. But in 100,000 years or so the sky as we know it today is likely to have a dramatically different aspect because of the the stars' proper motion, which was first recognized in Aldebaran. If we saw the sky after such a lapse of time, the constellations would have undergone such profound changes that their names would have lost all meaning.

Taurus

Star Map Series N II

The stars shown in series N II can be seen between the 20th and 40th degrees of northern latitude. The geographic map below shows the exact position of this latitudinal zone. These star maps therefore depict the skies that people living in the northern part of the world would see on vacation trips to the south. Here the constellations of Earth's Southern Hemisphere begin to become clearly visible above the horizon, constellations such as Centaurus (star maps N II/5 and 6) and Carina (star map N II/2).

The ecliptic, the apparent circle described by the Sun on the celestial sphere, is higher in the sky than in the maps of series N I. This means, too, that the planets look more dramatic and stay in the sky longer (see pages 16 and 113). It is also night earlier, especially in the southern parts of zone N II, because the Sun sets faster and dusk is brief.

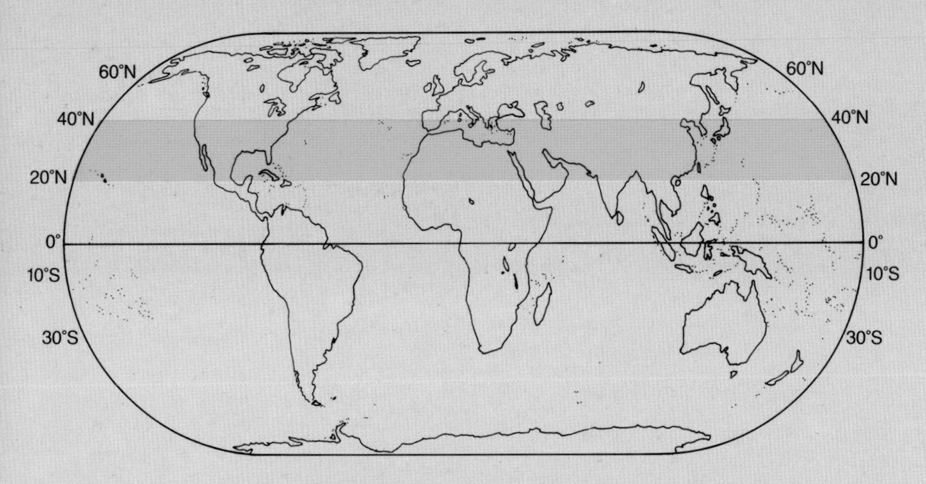

The star maps N II apply in the following countries and areas: the southern parts of the United States, Mexico, the Caribbean, Cuba, Spain, Portugal, southern Italy, Greece, Turkey, Israel, Egypt, Tunisia, Algeria, Morocco, Persia, Pakistan, Afghanistan, Saudi Arabia, India, Thailand, Burma, China, Korea, Japan, and Taiwan.

What the symbols in the star maps stand for:

-●- = Double stars: Pairs of stars that are particularly close to each other.
◉ = Variable stars: Stars that change in brightness.
□ = Nebulas: Clouds of gas and dust that shine in colors.
⁘ = Star clusters: Aggregations of many stars concentrated in one place.
O = Galaxies: Star systems (the Milky Way is one of many galaxies).

About the brightness and size of the stars marked in the maps:

0 = ✹ The brightest stars. They are easily visible even in big city skies.

1 = ● 2 = ● 3 = • 4 = • Scale of brightness 0 to 4. The smaller the number, the brighter the star.

◁ *The Great Nebula in Orion is a gas cloud that is 1,500 light-years distant from us.*

January

Star map N II/1

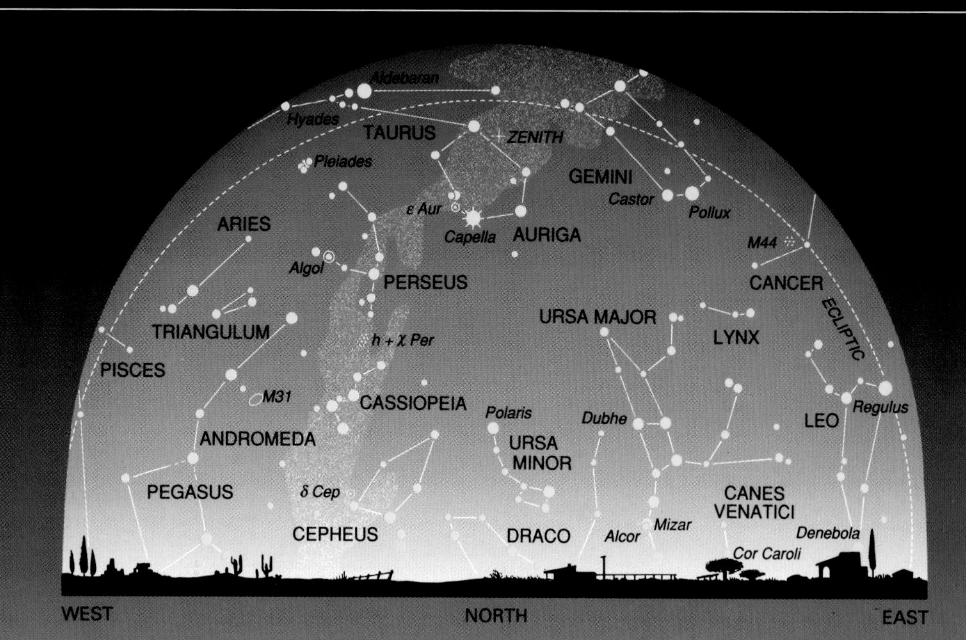

Today 88 constellations are recognized. It is easy to list them in order of size and position, but it is harder to say which constellation is the most magnificent. Still, most stargazers probably would agree that this rank should go to a constellation that shows up prominently high in the southern sky in January, namely **Orion**.

Orion is the only constellation that includes not just one but two of the sky's ten brightest stars: Betelgeuse (see page 131) and Rigel (see page 146). The rest of Orion's stars are also among the brighter ones and make up an impressive configuration in the sky. Three of them are arranged in a straight line or belt in the middle, with two stars each above and below the belt. From the dawn of human civilization, people who gazed up at the sky trying to interpret it thought they could recognize in this star pattern a gigantic figure: a hunter or a warrior, with Betelgeuse marking one shoulder and, in the lower half, Rigel together with its somewhat dimmer companion forming the knee or foot.

In keeping with his celestial grandeur, Orion is described in Greek mythology as an exceptionally mighty hunter. Usually he is depicted with his shield facing west to ward off the powerful bull (Taurus) and, with the other hand, swinging a club in the same direction. But the myths tell not only of Orion's contests with wild animals and human opponents but also of his interest in beautiful maidens. Thus he kept chasing after the Pleiades, the seven daughters of the giant Atlas, until they begged Zeus to save them from Orion's pursuits. Zeus responded by moving not only the Pleiades but Orion as well to the heavens, where the latter is forced to follow them without ever being able to catch up as they precede him across the sky from east to west. In another legend Orion is said to have made advances to Artemis, which also angered the gods. They sent a scorpion after him to sting him with its deadly poison. Just as Orion collapsed, Asclepius, the god of healing, happened to

January 1, 11 P.M.
January 15, 10 P.M.
January 31, 9 P.M.

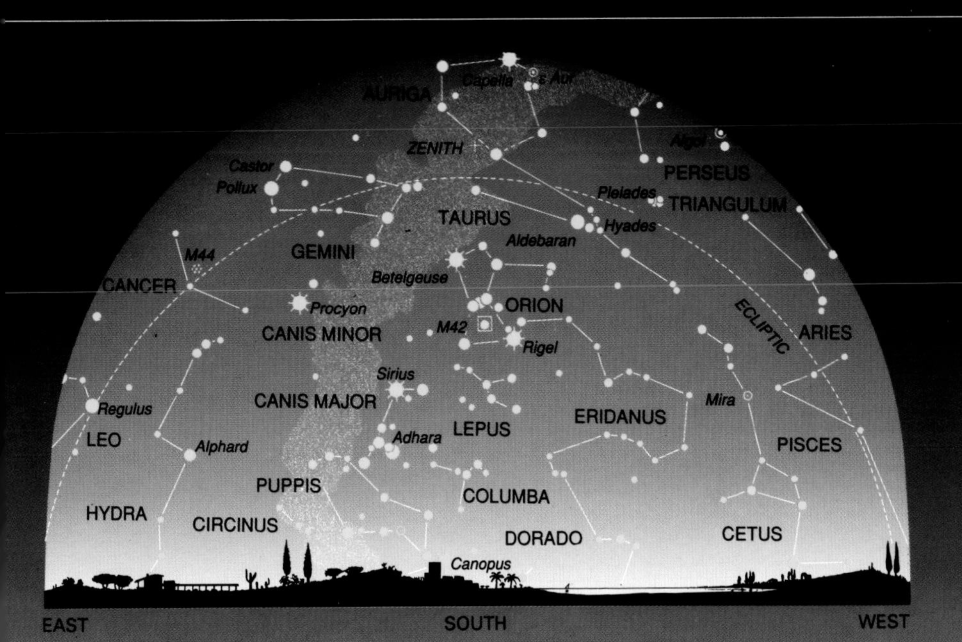

come by and tried to save Orion. But the gods' wrath was so great that Zeus hurled a bolt of lightning down at Orion as a final punishment. The lightning also killed Asclepius. All three earthly participants were turned into constellations. Asclepius appears in the sky under the name of Ophiuchus (the Serpent Bearer). The gods placed the constellations so ingeniously that the three never meet in the sky. Ophiuchus shows up there at a season when Orion becomes invisible (see stars in July, pages 66–67), and Scorpius is on the opposite side of the celestial sphere from Orion.

Orion

In addition to forming a striking pattern and having many bright stars, Orion also includes a number of other famous celestial objects, such as the Great Nebula in Orion, M 42. This is a gigantic, brightly shining gas cloud, which counts among the most colorful and notable celestial sights of this kind (see page 142).

February

Star map N II/2

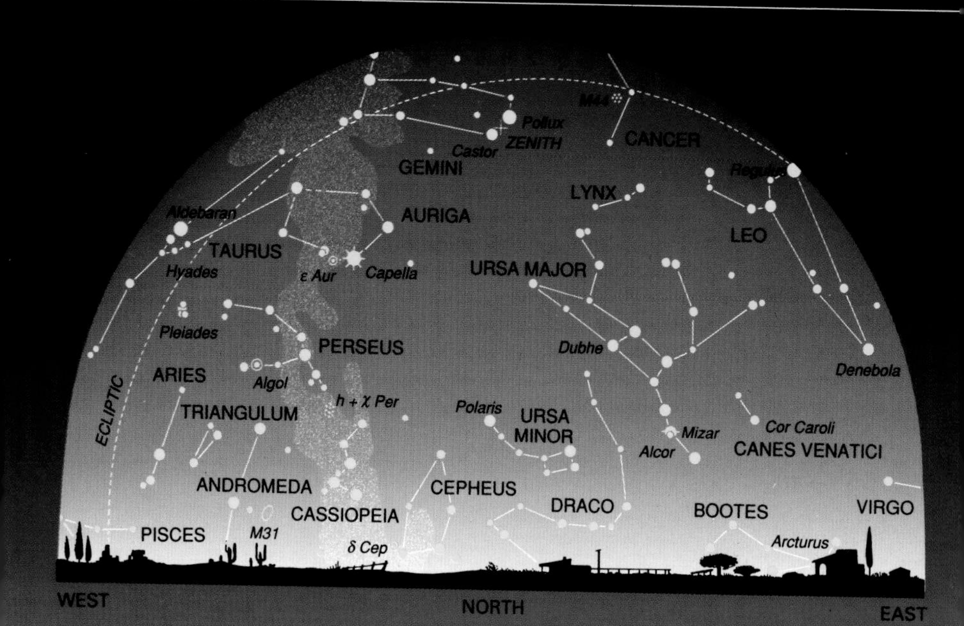

In keeping with the time of year, the typical winter constellations dominate the sky in February. Sirius, the brightest star in the sky, is in its highest possible position over the horizon, almost due south. Because of its outstanding brightness it always has been an object of greatest interest, and there is hardly a star around which as many stories are woven—and which has given rise to as many interesting scientific discoveries—as Sirius (see also pages 146–47).

Sirius is part of the constellation **Canis Major** (the Greater Dog). Somewhat higher in the sky we find Canis Minor (the Lesser Dog), which consists practically of one star only, Procyon, another one of the ten brightest stars in the sky (see page 145). In most star legends, the two constellations are regarded together as the dogs of the celestial hunter Orion. There is also, however, a story in which Canis Minor figures together with Boötes and Virgo (see pages 36–37). The expression dog days, which refers to the hottest period of

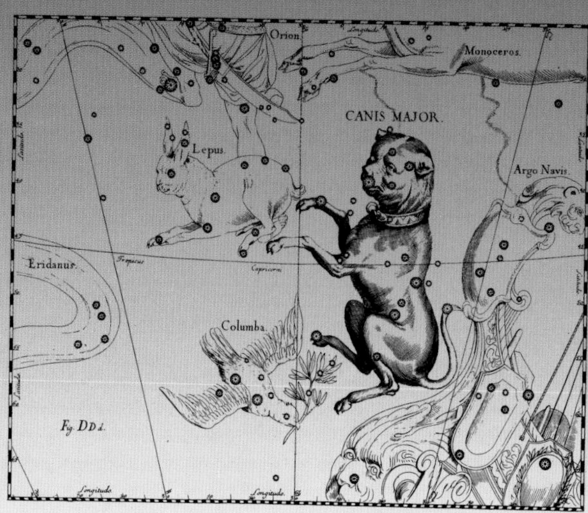

Canis Major

February 1, 11 P.M.
February 15, 10 P.M.
February 28, 9 P.M.

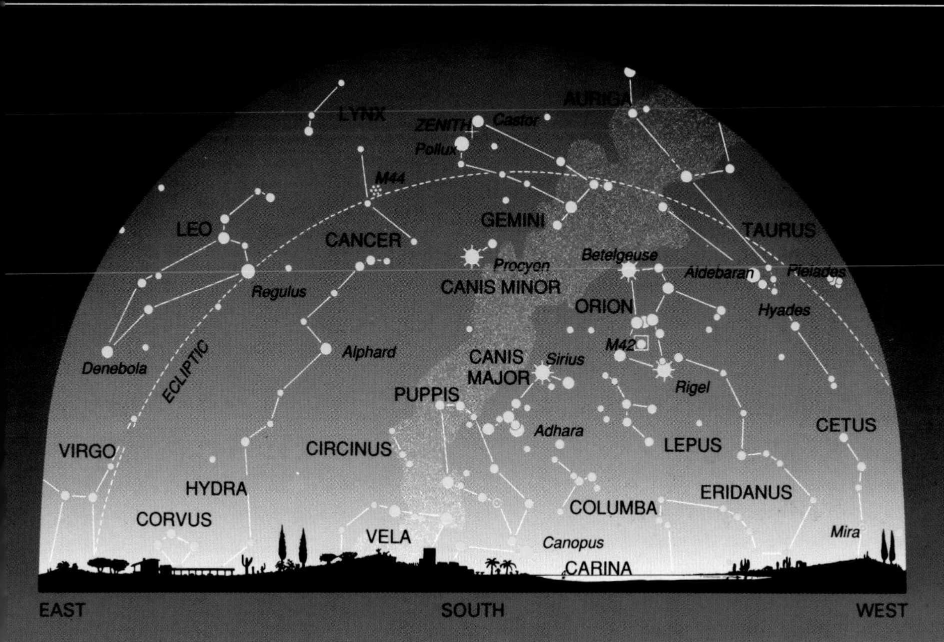

summer in August, derives from Canis Major. Canis Major is a winter constellation. In the months following February this constellation moves more and more westward in the evening until the Sun, traveling along the path of the ecliptic, is so close to it that it disappears from view as Canis Major accompanies the Sun across the daytime sky. This is the case in June and July. By August the Sun has moved farther along, and Canis Major makes its first brief appearance again in the dawn just before sunrise. After that it rises a little bit earlier every day, like all the stars, and in winter it can be seen in the evening sky again.

This first rise of Canis Major in the morning after a prolonged absence is more apparent than that of other constellations because Sirius, the brightest star, is easy to spot. That is why Sirius and Canis Major long ago became a symbol for the hottest days of summer, the dog days. Sirius even played a role of special importance in the calendar and daily life of ancient Egypt. The reason for this is that at the time around 3000 B.C. the first reappearance in the sky of Sirius coincided with the flooding of the Nile. These yearly floods were crucial for plentiful harvests, and thus Sirius, under the name of Sothis, was venerated as a god.

Sirius, which is one of the first stars to appear in the sky after sunset, serves as a lodestar even today. Next to it the other stars comprising Canis Major pale, even though they include a number of remarkable stars. Especially worth mentioning is Adhara, which is next in brightness within the constellation. It is 900 times more radiant than the Sun but is 500 light-years away. That is the only reason why it appears fainter to us than its bigger brother, Sirius, which is removed only 8.7 light-years from Earth.

March

Star map N II/3

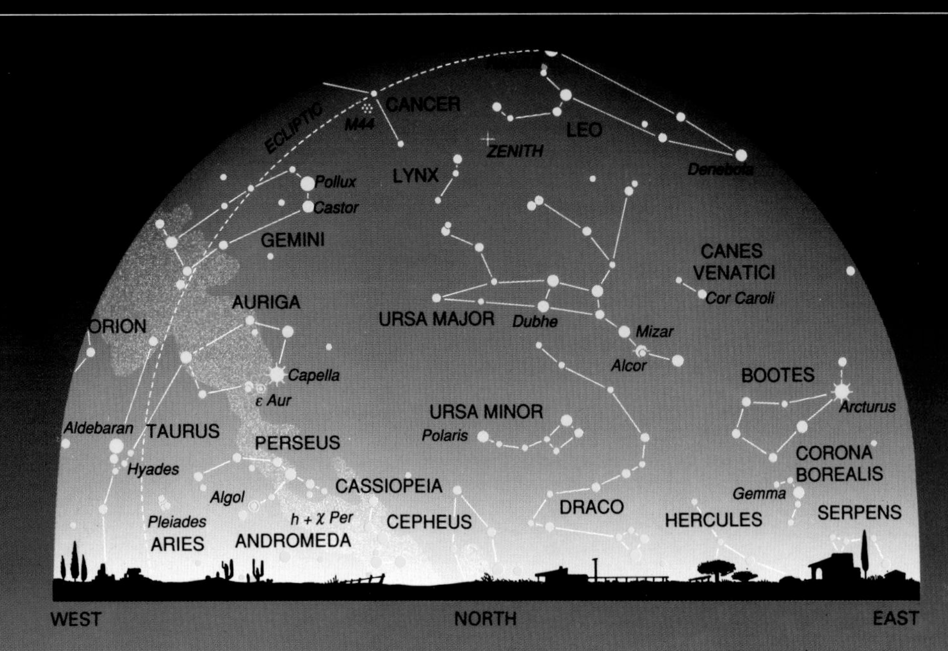

Beginning in March, the month that ushers in spring, the winter constellations are in the western sky in the evening. They sink below the horizon around midnight. This is the last time until the following winter that the Winter Hexagon can be seen, this hexagon in which are gathered six stars belonging to some of the most beautiful constellations: Capella in Auriga; Aldebaran in Taurus; Rigel in Orion; Sirius in Canis Major; Procyon in Canis Minor; and, finally, Castor or Pollux in Gemini.

It is not altogether clear which of the twins should be included in the hexagon because they are more or less equally luminous and stand so close to each other that the name **Gemini** (the Twins) comes to mind quite automatically. In Greek mythology, from which the names Castor and Pollux come, the twins have different fathers. The story is that Zeus became infatuated with beautiful Leda, who loved her husband truly. So Zeus changed himself into a swan and visited her. Leda, who suspected nothing, made love with her husband that same night, and thus she came to bear male twins who had different fathers. Pollux was the son of Zeus and therefore immortal, whereas his brother, Castor, was an ordinary mortal. The brothers loved each other dearly and performed many heroic deeds together. Eventually Castor died, and Pollux asked Zeus to make him mortal, too. Zeus refused to grant this wish, and so Pollux decided to follow his brother into the underworld. Deeply impressed by such profound brother love, Zeus moved both of them into the heavens. Modern astronomers view these two stars in a very different light with no sign of brotherhood between them. Pollux is 35 light-years away from us, Castor, 50 light-years. Pollux is a single star, while Castor is made up of six stars that form one of the most unusual multiple star systems in the sky.

Gemini is one of the zodiacal constellations through which the Sun progresses. When it reaches Gemini, the Sun is in its highest position

March 1, 11 P.M.
March 15, 10 P.M.
March 31, 9 P.M.

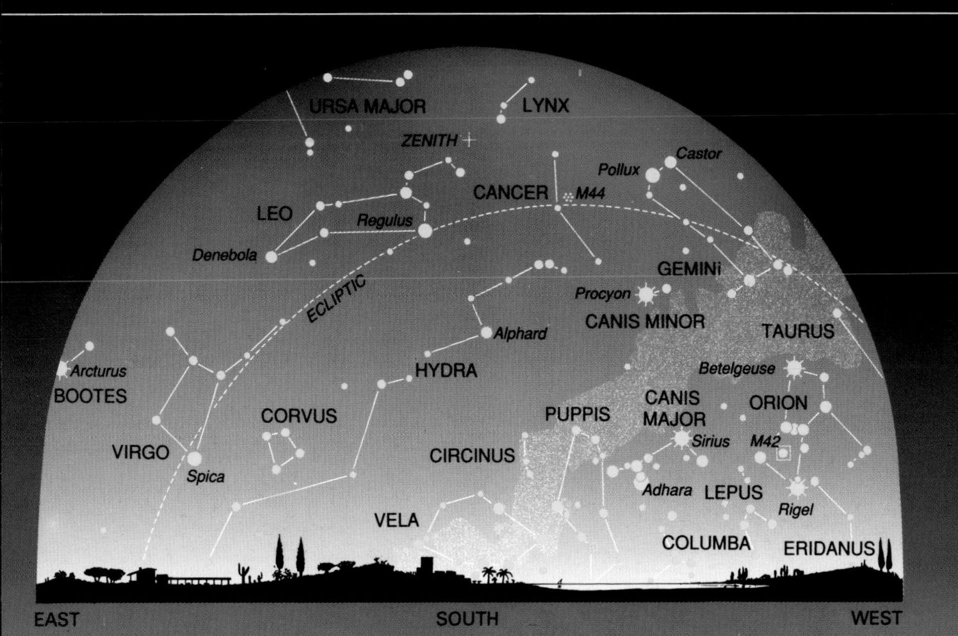

in the sky as seen from Earth's Northern Hemisphere. The fact that Gemini belongs to the zodiacal constellations is also the reason why two of the three major planets that are visible only through a telescope were first discovered in this constellation. In 1781 the astronomer William Herschel, observing the sky from England, first saw Uranus, and in 1930 the American Clyde Tombaugh discovered Pluto, also in Gemini.

Cancer (the Crab), also a zodiacal constellation, is next to Gemini to the east. It and Libra are the least conspicuous constellations of the zodiac. Cancer is said to have been crushed underfoot in the course of Hercules's battle with Hydra (the Water Snake) (see pages 62–63). The luckless crab nevertheless was raised to the heavens by the mother goddess Hera. Cancer's claim to fame is based primarily on the open star cluster M 44, which is one of the brightest in the sky and often is referred to as the Praesepe (Latin for manger) or the Beehive (see page 145).

Gemini

April
Star map N II/4

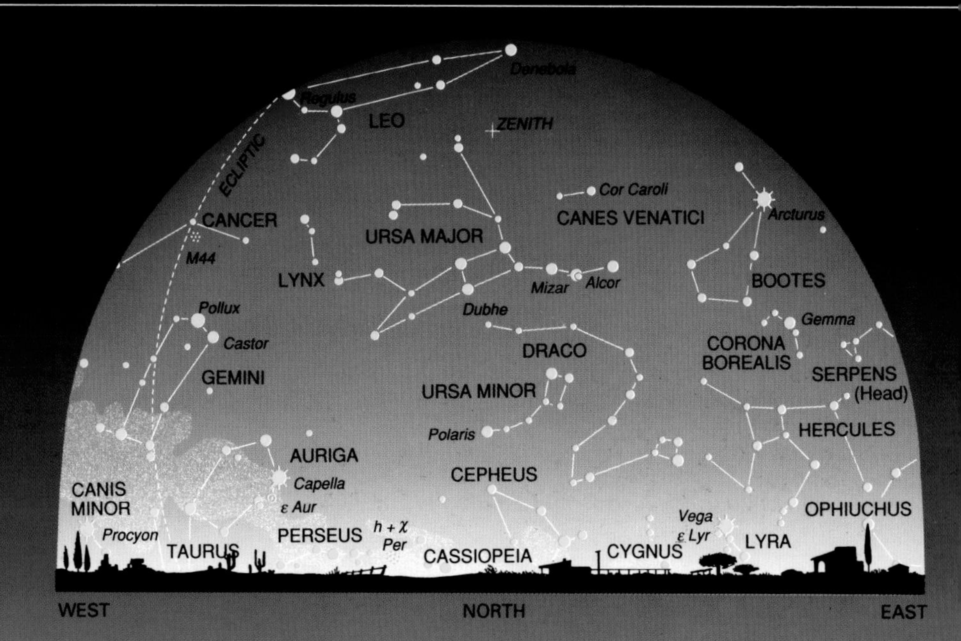

To get oriented in the night sky, many observers look not only for the constellations shown in the sky charts but also resort to other star combinations to help them get their bearings. Thus there is a Winter Hexagon, an Autumn Square (the Great Square of Pegasus), a Summer Triangle, and a Vernal Triangle to look for. The Vernal Triangle stands especially high and prominent in the southern sky during April, the first month that belongs entirely to spring. The triangle is made up of the brightest stars in the constellations Boötes, Virgo, and Leo.

The brightest star in Leo is named Regulus, which means little king. **Leo** (the Lion) is one of the constellations that is easy to spot in the sky because it is made up of two easy to recognize trapezoids. The lower, larger trapezoid has Regulus and the next brightest star, Denebola, as its bottom corners, and the smaller trapezoid sits atop the bigger one to one side. It doesn't take a great deal of imagination to see the head and body of a mighty lion in this arrangement of stars, and it is not surprising that the name Leo, as applied to this area of the sky, is of great antiquity. Leo is associated with one of the great feats of Hercules, whose constellation appears in the April sky, too, in the east. Hercules's first labor was to slay the Nemean lion, a fearful beast whose skin was immune to all injury, so that Hercules could not subdue the animal with either bow and arrow or a club. He therefore grabbed it by the throat and choked it to death. Then he skinned the dead lion and wore its pelt as a fear-inspiring garment. But because the Nemean lion was an immortal animal, the gods brought it back to life as a constellation in the heavens.

Leo is one of the zodiacal constellations along the ecliptic. Regulus, its brightest star, stands almost on the exact line of the ecliptic and thus is obscured periodically by the Moon. It is even possible for planets to come between it and Earth, so that its light rays are blocked. But because the planets appear

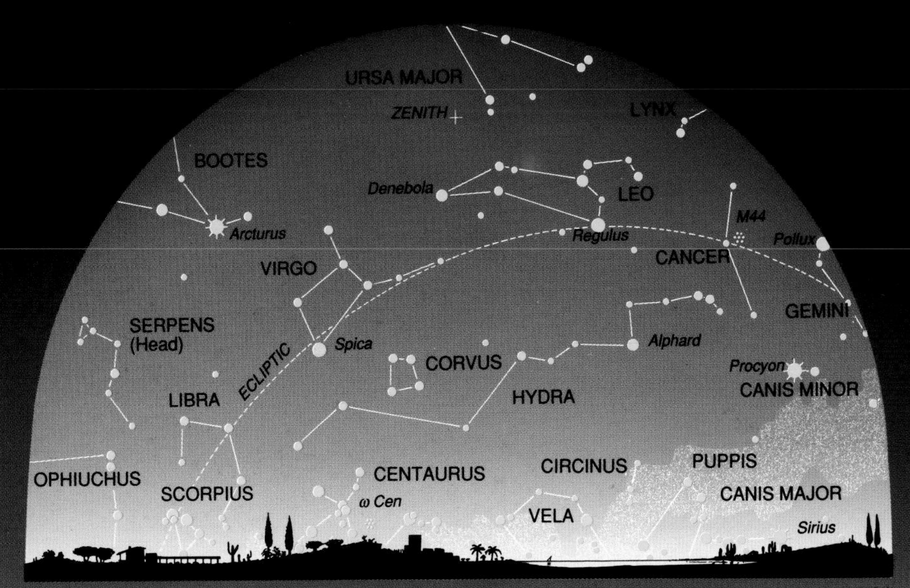

so much smaller on Earth than does the broad face of the Moon, this happens only very infrequently. The only time this kind of celestial event has been observed since the beginning of modern astronomy was on July 7, 1959, when Venus covered up Regulus. Regulus is 80 light-years away from Earth and is 160 times more radiant than the Sun. Its diameter is estimated to be five times that of the Sun.

The second brightest star in Leo, Denebola, further reaffirms the name given by astronomers to this region of the sky. It comes from the Arabic and means lion's tail. Denebola is of medium size. It is 40 light-years away and about 20 times brighter than the Sun.

Leo

May

Star map N II/5

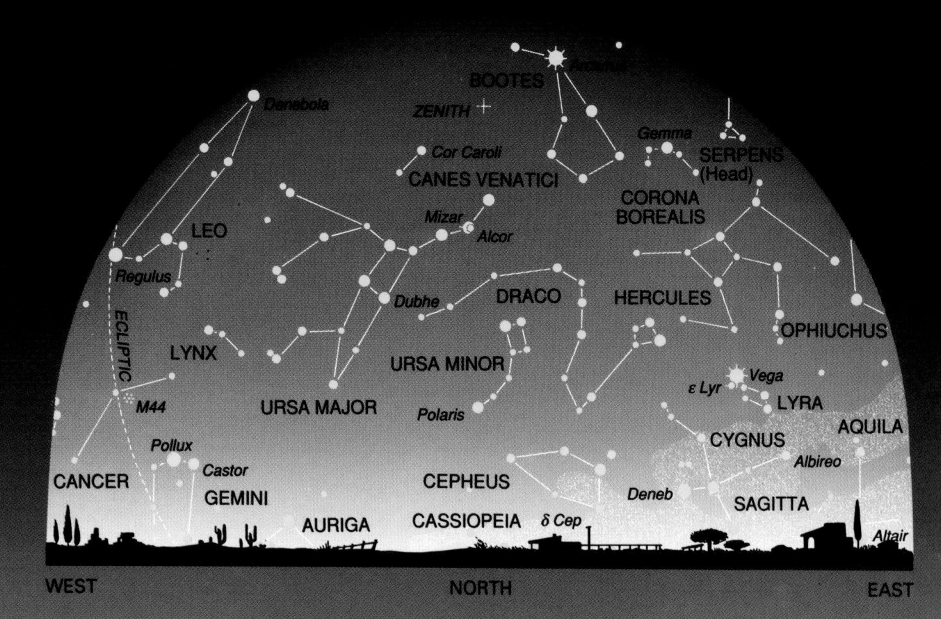

Because the Sun sets later at this time of year, the visibility of the stars is noticeably lessened. Dusk lasts longer, so it is not until about 10 P.M. (11 P.M. where daylight saving time is in effect) that the constellations can be made out.

The paucity of bright stars is particularly striking in the lower half of the southern sky, where a surprisingly large celestial expanse contains only one constellation, Hydra. **Hydra** (the Water Snake) is the largest of the 88 officially recognized constellations, but its stars are much fainter than those of the smallest constellation, the Southern Cross, which is invisible from the 30th degree of northern latitude. The long drawn out body of Hydra has only one fairly bright star, Alphard. The name is fitting in view of the faintness of Hydra. It translates as the solitary one, that is, it is the only star to draw the eye's attention in an otherwise starless region of the sky. Alphard shines with a slightly reddish light and is 90 light-years away from Earth.

Hydra is connected to the myths about Hercules and his famous 12 labors. After Hercules had vanquished the lion, whose constellation is also visible in May in the southwestern sky, his next task was to slay the Hydra. This monster was especially terrifying because it had seven heads, and for every head that was cut off, new ones grew back instantly in its place. Thus Hercules found himself in serious straits until Athene helped him with some advice. On her suggestion he now asked his nephew, who had accompanied him, to hand him flaming torches with which he cauterized each stump after a head was severed and thus prevented the sprouting of new ones. When Hera, Hercules's arch enemy, saw that he was about to triumph after all, she sent the crab after him at the last moment to bite him, but Hercules crushed it underfoot, whereupon it was turned into the constellation Cancer.

On a clear night, once it is dark enough, the whole body of the Hydra can be made out. The brighter stars, which

May 1, 11 P.M.
May 15, 10 P.M.
May 31, 9 P.M.

Add one hour if daylight saving time is in effect.

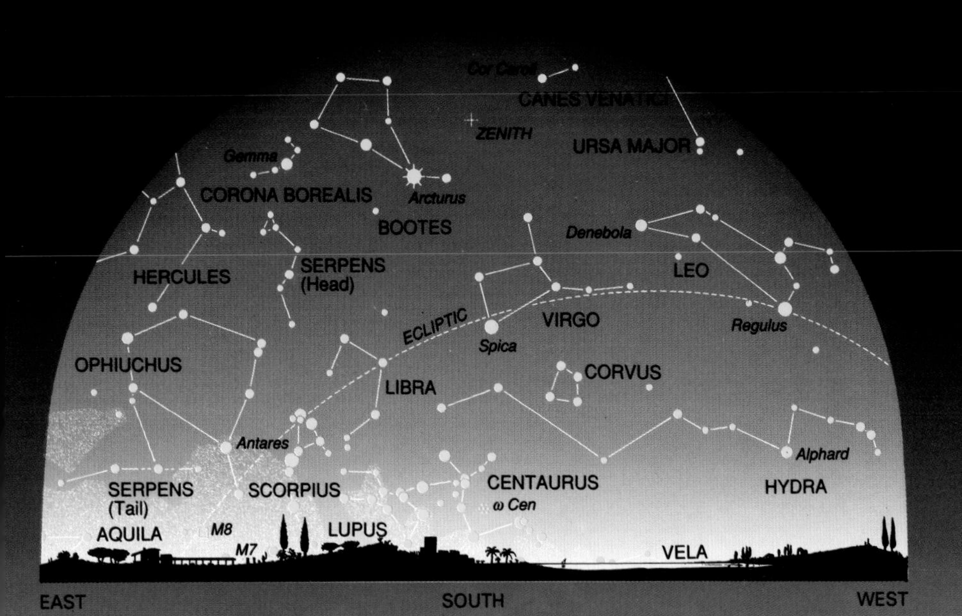

are shown in the chart, form a snakelike line across the sky. The head, indicated by two somewhat brighter stars than most of the rest, is next to Cancer. Many other constellations show up more faintly in the May sky in late evening. In the north there are Cepheus and Lynx. The lynx, about which there is not much to report, was defined by Johannes Hevelius, who gave it its present name because, supposedly, one had to have "the eyes of a lynx" to make out the faint stars of this constellation.

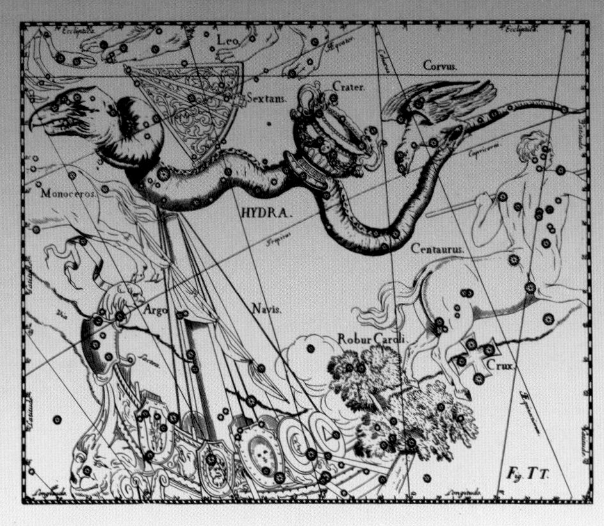

Hydra

June

Star map N II/6

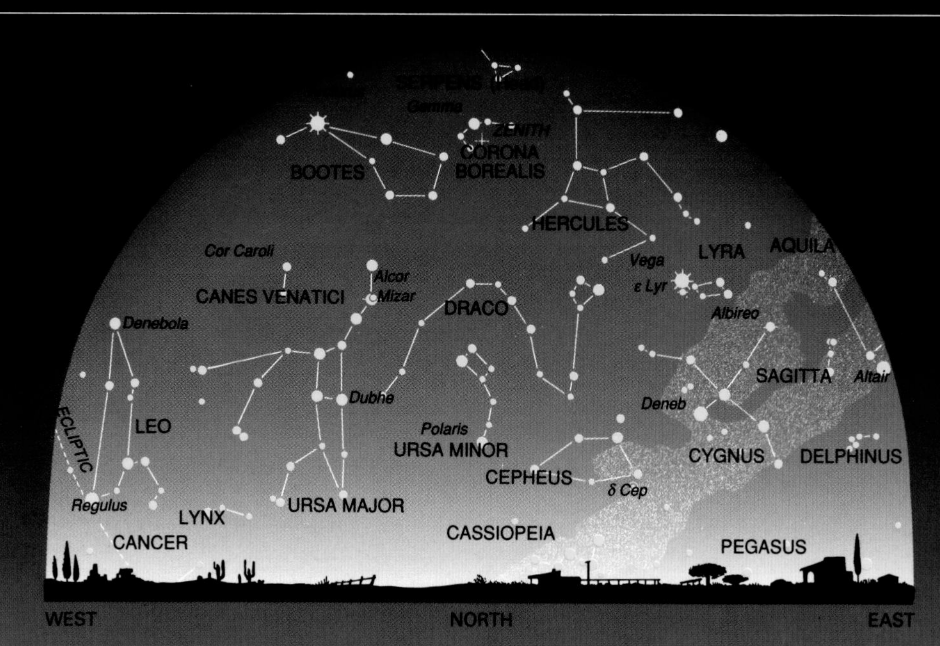

June marks the beginning of summer. Everywhere in the Northern Hemisphere the Sun now stands highest in the sky; the days are longer than at any other time of the year, and the nights are shorter, although all this is not as pronounced between the 30th and 40th degrees of northern latitude as it is farther north (see What Causes the Seasons of the Year? on page 13). Just like May, June is a month of rather inconspicuous constellations that are seldom mentioned but that nevertheless display a number of interesting features.

Due south and on the ecliptic we find Libra. **Libra** (the Scales) is the only constellation of the zodiac that is not named after an animal or a person. The name Libra has direct reference to the movement of the Sun. Originally this constellation was considered part of Scorpius, with the two brightest stars making up the claws of the scorpion. But very early, and probably originating in Egypt, the name Libra gained currency because the Sun passes through Libra at the

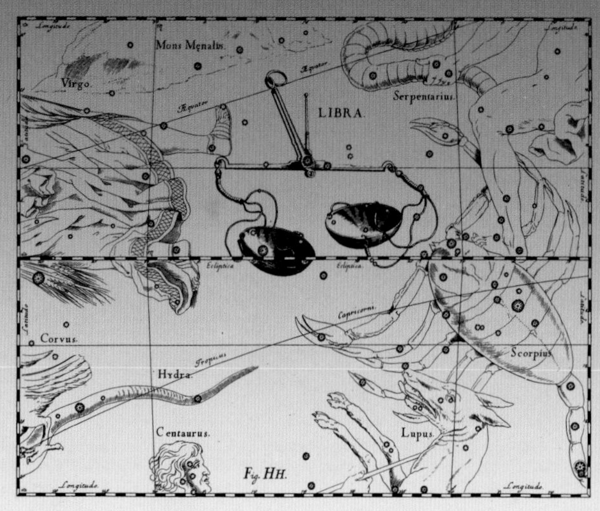

Libra

June 1, 11 P.M.
June 15, 10 P.M.
June 30, 9 P.M.

Add one hour if daylight saving time is in effect.

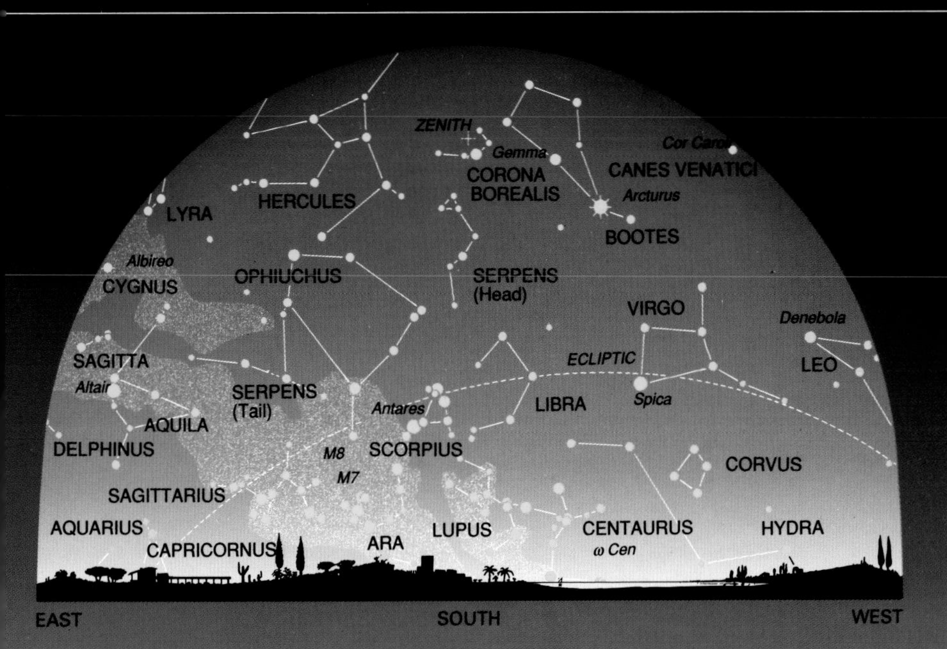

time of the fall equinox in September, when day and night are of equal length. After the equinox, the Sun begins to sink lower in the sky until winter begins. The scales, which stand as a symbol for balance and justice, thus seemed an obvious candidate to mark this special position of the Sun in the sky, especially the equality of day and night in the fall, when winter is still distant but the hot days of summer are clearly past. As is the case with the other constellations of the zodiac, planets frequently show up in Libra (see The Planets Appear in the Constellations of the Zodiac on page 113).

Next to Libra on the ecliptic, to the east, is Scorpius, which also belongs to the zodiac. On the west, Libra is preceded by Virgo and Leo. South of Virgo is the constellation Corvus (the Crow), whose origins are obscure. We do know that it goes back to the Greeks, but the legends associated with it are contradictory. According to one of them, the crow was sent by Apollo to fetch water from a spring. But the crow found some delicious figs on the way and stopped to eat them. When he returned, he was clutching the water snake in his claws and claimed he had found no water because the snake had drunk the spring empty. Apollo, realizing immediately that this was a lie, banished the crow to the heavens, along with the Hydra, where he is condemned to suffer eternal thirst.

Because in June the sky is so bright before midnight, it is especially important to remember how to get one's bearings by first finding the Big Dipper in Ursa Major. Ursa Major is in the northeast during June. Once the characteristic pattern of the Big Dipper is identified it is not difficult to find Boötes, Virgo, and Cassiopeia, as well as many other constellations.

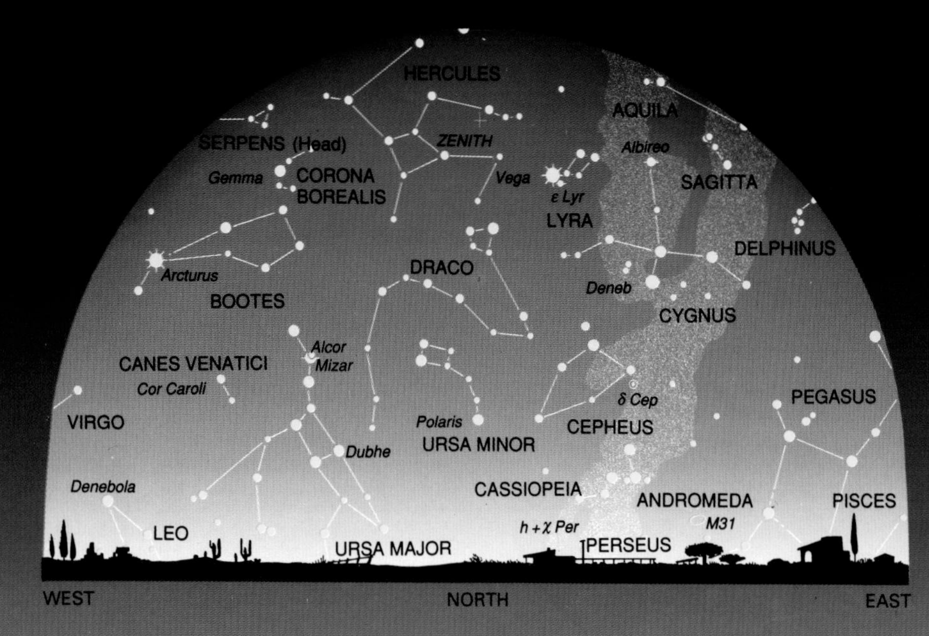

In July, the distance between the Sun and Earth grows largest. We have hot summer weather in spite of this because of the tilt of Earth's axis, which results in the Sun passing through the sky in a high arc during the day and thus shining longer (see pages 12–13). The ecliptic, the apparent path of the Sun across the celestial sphere, is consequently very low in the southern sky during July evenings and nights, and whatever planets appear are close to the horizon, too.

But the southern part of the sky is particularly interesting on account of the constellations that show up there. Sagittarius, Scorpius, and Ophiuchus are all there, and the Milky Way passes through them, being brightest in Sagittarius because the center of the galaxy lies in that direction (see pages 92–93). All three constellations mentioned are part of the zodiac, but **Ophiuchus** (the Serpent Bearer) is the one that requires comment because it is the thirteenth zodiacal constellation. In the classical zodiac that is still used today in horoscopes, there is no serpent bearer. Ophiuchus was not included among the zodiacal constellations until 1930, when the boundaries of the constellations were revised (see page 16).

The serpent bearer is closely linked to **Serpens** (the Snake or Serpent). In old representations of the starry skies, Ophiuchus is shown holding the snake in his hands, and the modern redefinition of the constellations still adheres to this old interpretation. Serpens is the only constellation that consists of two separate parts, namely the head and the tail. This means that though there are 88 constellations, the sky is divided into 89 sections because the snake consists of two unconnected segments.

The snake and the snake bearer are associated in Greek mythology with the god of healing, Asclepius. Even today a snake wound around a staff is the symbol of medicine and pharmacology. Asclepius was a renowned physician. In fact, he practiced his art so suc-

July 1, 11 P.M.
July 15, 10 P.M.
July 31, 9 P.M.

Add one hour if daylight saving time is in effect.

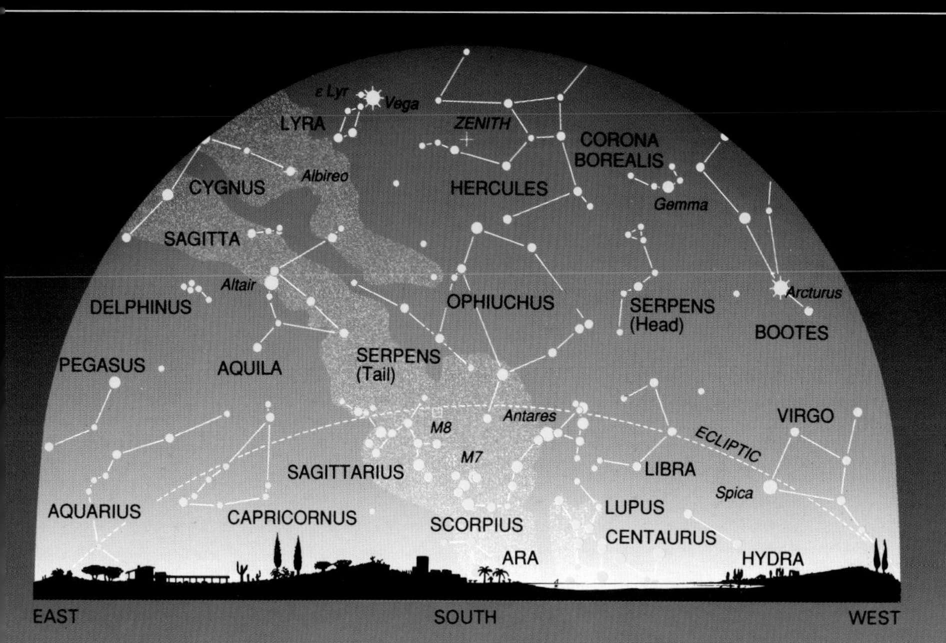

cessfully that he eventually aroused the ire of Pluto, the god of the underworld, who complained to Zeus that not enough mortals were arriving in his realm anymore. Zeus thereupon struck Asclepius with a bolt of lightning. In another tale Asclepius tried to save the mortally wounded Orion, whereupon he suffered the same fate.

In 1604 Ophiuchus was the site of the eruption of a supernova. This was the last such blaze that could be seen with the naked eye from northern latitudes. One astronomer who watched with special closeness was Johannes Kepler, and it became known as Kepler's Star. This supernova in Ophiuchus was visible for 18 months and both fascinated and alarmed the people of the early seventeenth century because only 30 years earlier another supernova had erupted in Cassiopeia (which can be found in the northern sky in July). At the height of its blaze, the supernova in Ophiuchus was brighter than Jupiter (see pages 149–50).

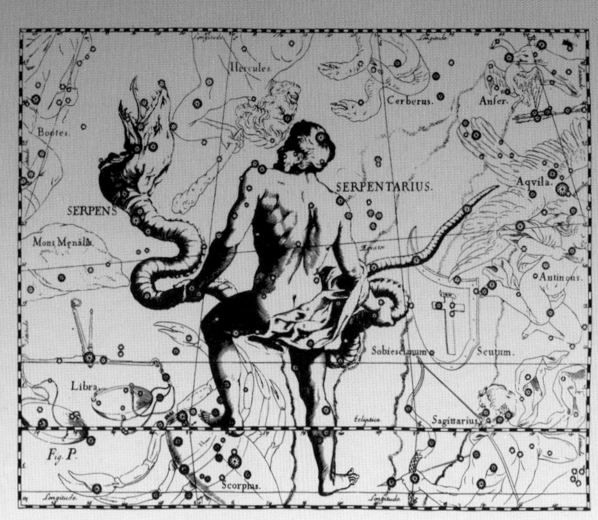

Ophiuchus and Serpens

Star map N II/8

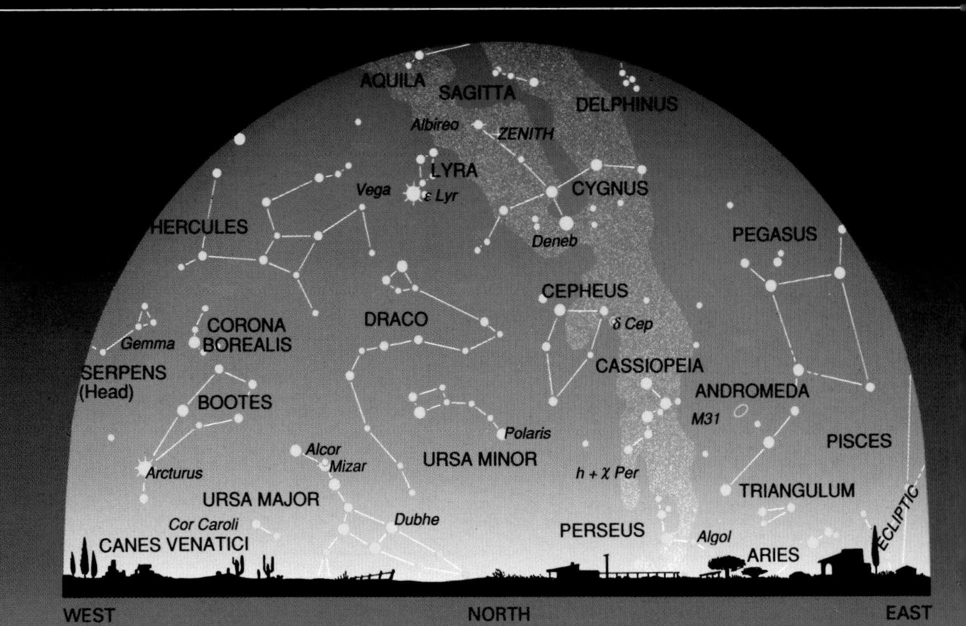

During the hottest month in Earth's Northern Hemisphere, the Summer Triangle stands out clearly overhead. It is made up of the brightest stars of Cygnus, Lyra, and Aquila, namely Deneb, Vega, and Altair. Low on the northeastern horizon, Perseus rises with its double cluster h and χ (chi) and the famous variable star Algol.

Perseus is also the site, or radiant, of the most spectacular meteor shower of the year, the Perseids. Around August 10, large numbers of meteors seem to stream from this constellation. At this time of year, Earth traverses an exceptionally dense accumulation of dust particles that enter Earth's atmosphere where they heat up the air and create the familiar light streaks we call meteors or shooting stars (see page 138). The best time to watch for the Perseids is after midnight because Perseus stands higher in the sky then. The Moon should interfere as little as possible and therefore not be in its full phase (see table on pages 106–107).

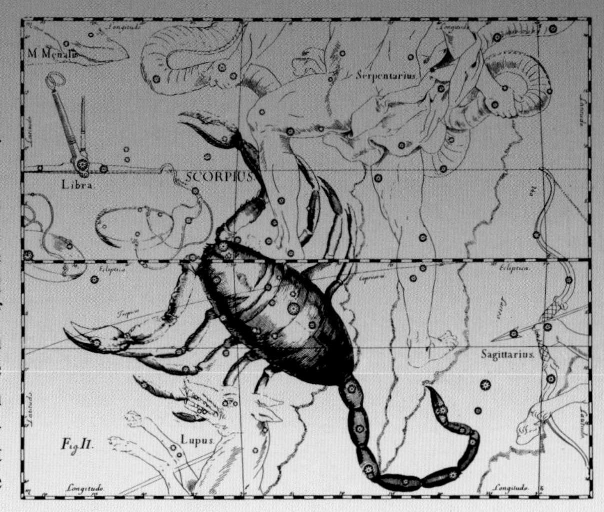

Scorpius

August 1, 11 P.M.
August 15, 10 P.M.
August 31, 9 P.M.

Add one hour if daylight saving time is in effect.

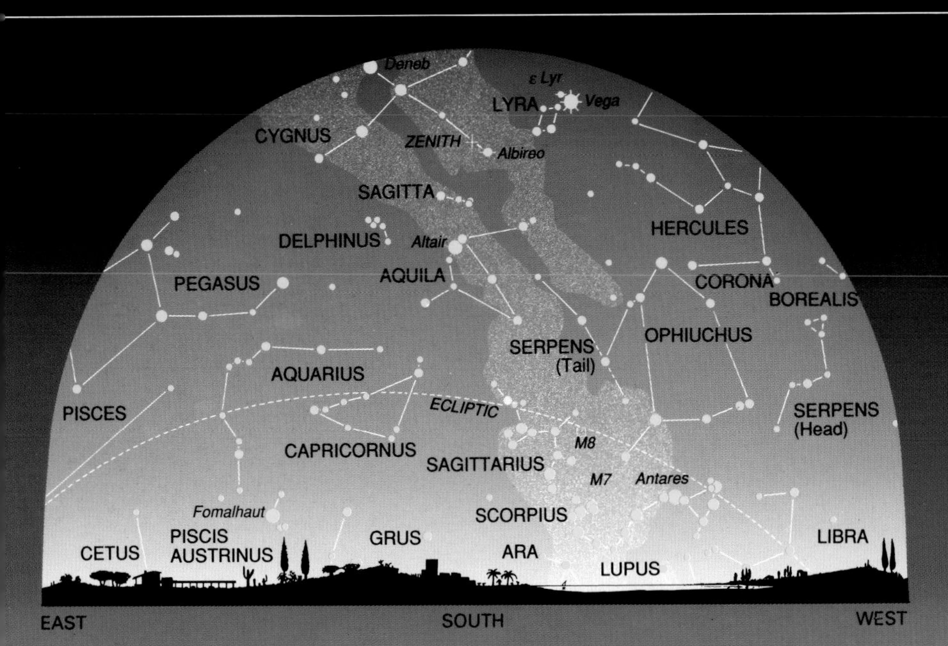

In the south, the Milky Way is prominently visible because it rises almost vertically from the horizon to the zenith. Along the way it passes through an interesting constellation, namely **Scorpius**, which is one of the zodiacal constellations (see page 16). Scorpius, too, is of ancient origin, and it includes a number of resplendent stars, including Antares, the brightest in this constellation. Antares is one of the few stars that shows a clear color. It shines with a reddish light, to which it owes its name. Because Mars is red—like Antares—and because, as a planet, Mars also moves along the ecliptic and therefore appears in Antares's vicinity with some frequency, it was dubbed anti-Mars, for Ares was the Greek god of war, known to the Romans as Mars.

Antares is so close to the ecliptic that it often is hidden by the Moon. It is an unusual star about 500 light-years away. Astronomers term it a supergiant—like Betelgeuse in Orion (see pages 131 and 55). Its diameter exceeds that of our Sun by 700 times. If it were located where the Sun is, it would extend beyond the orbit of Mars and swallow Earth up completely. Its light is at least 7,600 times brighter than that of the Sun. Also in Scorpius the open star cluster M 7 is found, one of the few clusters visible to the naked eye (see pages 147–49).

The story of Scorpius is closely linked to those of some of the other constellations. Thus, according to one myth, Scorpius mortally stung Orion (see page 55) and was placed in the heavens in such a way that it and Orion are always on opposite sides of the sky and consequently never meet. Scorpius is a summer constellation, whereas Orion appears only in winter.

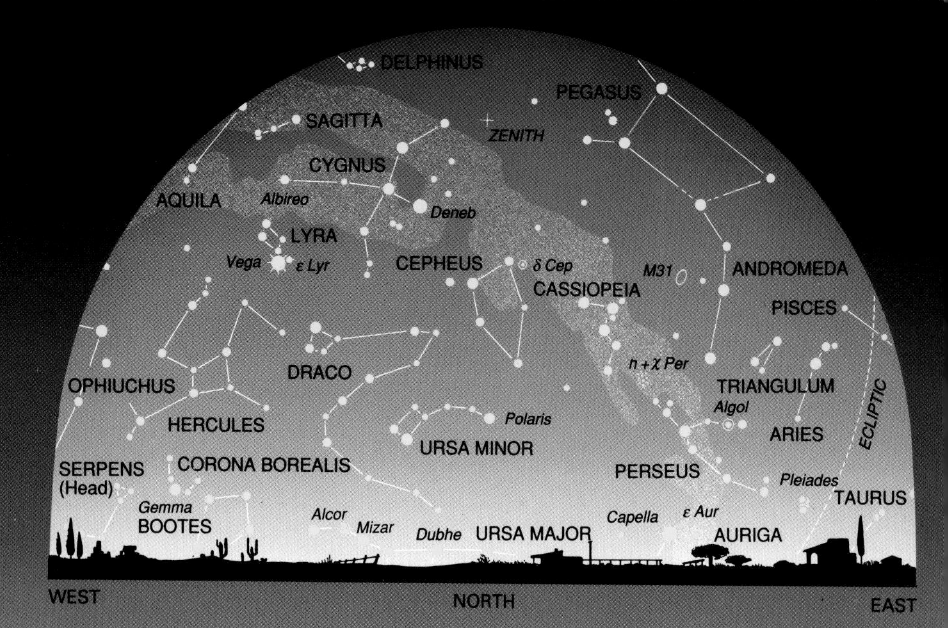

September marks the beginning of fall. Day and night are of approximately equal length (see pages 12–13 and 104–05), so that the stars are again easier to watch because they show up earlier in the evening.

This is important because now many of the familiar constellations are low on the horizon in the evening and only relatively faint stars occupy the better positions in the sky. Ursa Major, for instance, is extremely low in the north—so low, in fact, that in this southerly latitude some of its stars have sunk below the horizon. The ecliptic is about half way up in the sky in the south, where it transects Capricornus and **Aquarius** (the Water Bearer). Both these constellations consequently often form the backdrop for planets, which also can be observed easily in September. In September of 1846, the planet Neptune, one of the three planets that are not visible to the naked eye, was first detected in the constellation Capricornus.

Capricornus and Aquarius lend themselves particularly well to illustrate an important phenomenon in the movement of the stars. On many geographical maps, the term Tropic of Capricorn is used even today to indicate the parallel of latitude at 23½ degrees south of the equator. Every year when the Sun is at this southern limit of its annual path, that is, at the beginning of winter in the Northern Hemisphere, it appears exactly on the zenith at noon, that is, directly overhead. In the time before Christ's birth when these ancient concepts were formed, the Sun stood among the stars of Capricornus at this time of year. This no longer is true today; it now stands in the constellation Sagittarius at the beginning of winter, and it therefore would be more accurate now to speak of a Tropic of Sagittarius.

This shift is due to the same phenomenon that also accounts for the changing position of the north celestial pole (see the constellation Draco, on pages 42–43). It is called precession, that is, the gradual moving ahead of the most important lines and

September 1, 11 P.M.
September 15, 10 P.M.
September 30, 9 P.M.

Add one hour if daylight saving time is in effect.

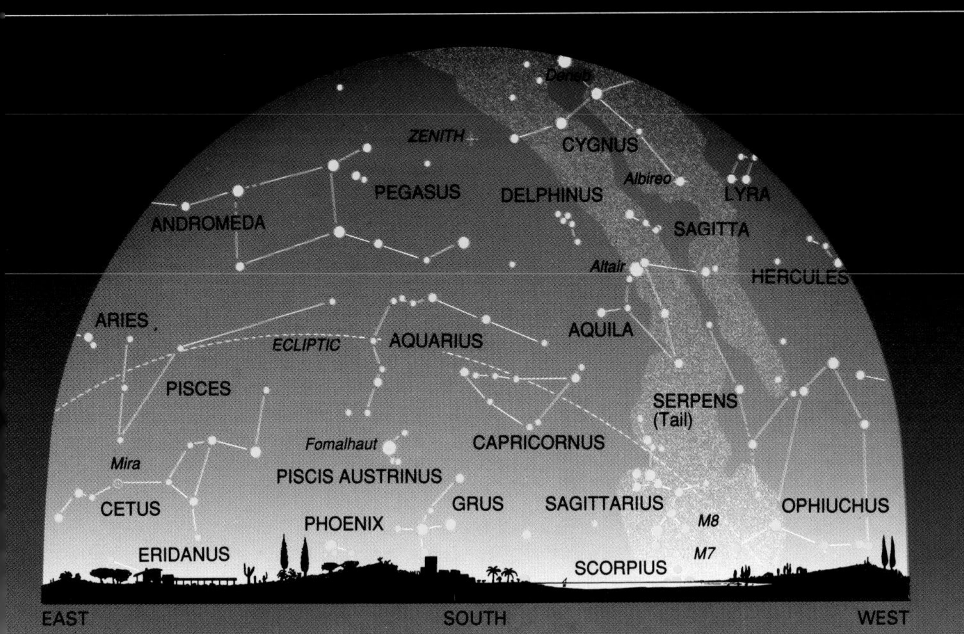

points of orientation—the celestial equator and the celestial poles—against the background of the stars. Once in the course of 26,000 years these lines and points describe a complete circle in the celestial sphere. Since the naming of the Tropic of Capricorn, the lowest point the Sun reaches on the ecliptic has moved from Capricornus to Sagittarius; and the spot where it always stands at the beginning of spring, known as the vernal equinox, has moved similarly out of Aries and into Pisces. Because of this new position, astrology calls the present time the Age of Pisces. But because the vernal equinox keeps moving along the ecliptic, it will enter the constellation Aquarius around the year 2400. Then the Age of Aquarius will begin, when—the story goes—everything on Earth will be better and more peaceful. This is the astronomical background of the popular song "The Age of Aquarius" from the world famous musical *Hair*.

Aquarius

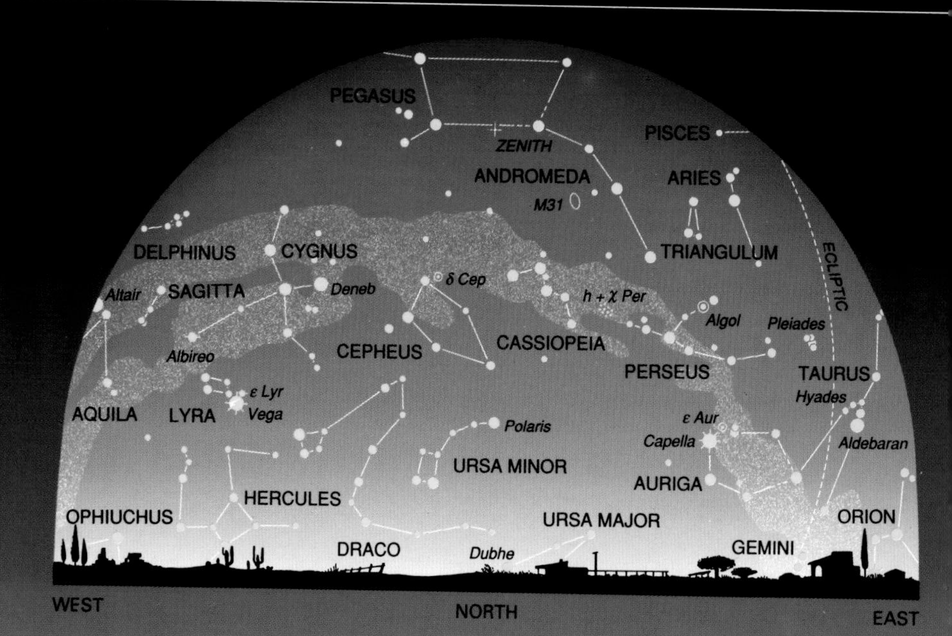

Every season of the year has its typical constellations. Because, in its apparent motion, the Sun completes a circle around the celestial sphere in the course of a year, different stars are seen at night from month to month, stars that six months later cross the daytime sky along with the Sun, invisible because of the bright sunlight. It is thus the Sun that determines, with its apparent movement, which stars are visible during which part of the year, whereas Earth's rotation around its own axis accounts for the fact that individual stars are visible in different places at different hours of the night (see The Movement of Earth on page 8).

The night sky therefore looks the same at a given season every year, except for the position of the planets (see pages 113–17). In the fall, the Autumn Square, also called the Great Square of Pegasus, is particularly conspicuous. It is formed by four stars from the constellations **Pegasus** and Andromeda and stands directly overhead in October, in the highest point of the sky, the zenith.

The constellation of Pegasus usually is represented by this almost perfect square, even though the star forming one of its corners is part of Andromeda. This usually is indicated in star maps by dotted lines. Pegasus is named after a famous legendary beast of Greek mythology, a winged horse. Pegasus was born from the blood of the Gorgon Medusa, who was so terrible to behold that anyone who laid eyes on her instantly turned to stone. When Perseus set out to conquer her, he had the foresight not to look at her directly but only at her reflection on his shield. In this way he was able to overcome her. When he finally killed her, Pegasus sprang forth from the severed head, spread his wings, and flew up to the gods in the heavens. Perseus, Andromeda, Cepheus, and Cassiopeia all figure together in Greek mythology and are all visible in the northern sky during October.

Pegasus went on to perform many heroic deeds at

October 1, 11 P.M.
October 15, 10 P.M.
October 31, 9 P.M.

Add one hour if daylight saving time is in effect.

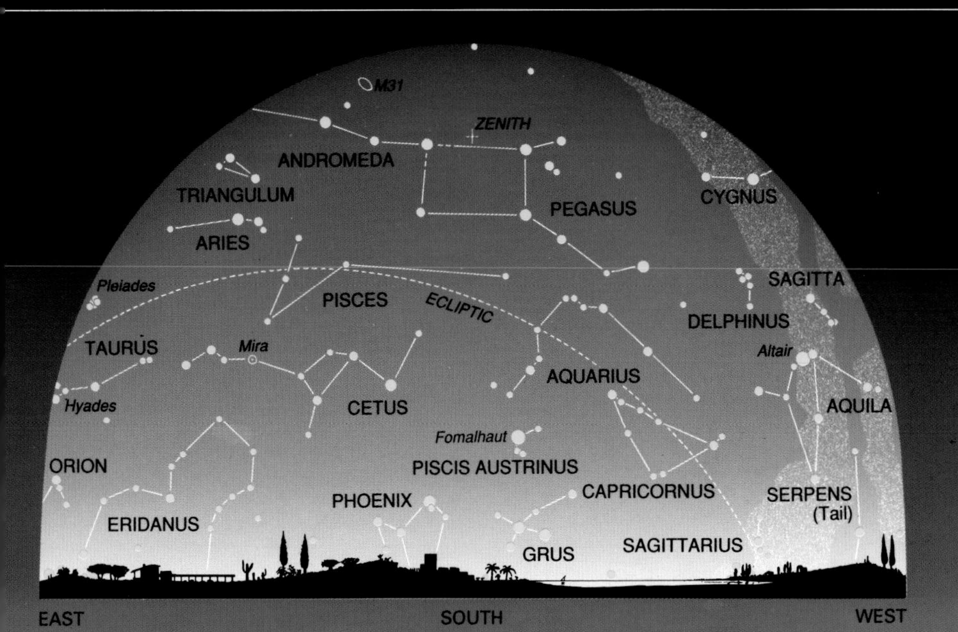

the behest of the gods. At one point, when he landed on the island Helion, he struck the ground with his hooves and caused the fountain of wisdom to flow. Anyone who drank from this spring was filled with poetic imagination, so that Pegasus later became a symbol of poetry.

The constellation Pegasus extends far beyond its Great Square, reaching as far as Cygnus and Delphinus in the west. In spite of its size, Pegasus contains very few remarkable stars or unusual celestial objects. The opposite is true of Andromeda, which displays the famous Andromeda Galaxy, one of the few galaxies visible to the naked eye (see page 135). Its official designation is M 31, which means that it is the thirty-first entry in the catalog compiled by Charles Messier in the eighteenth century. Altogether, the view into the north yields more during October than does the southern sky. An exception is Piscis Austrinus (the Southern Fish), which contains one bright star, Fomalhaut (see pages 94–95).

Pegasus

November

Star map N II/11

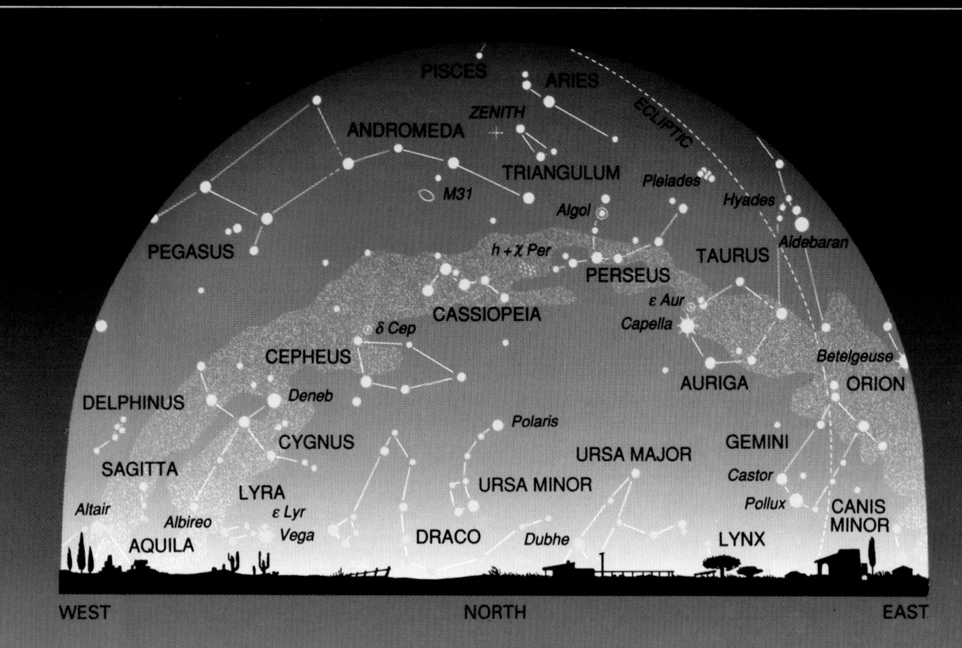

A great many constellations go back to the same cycle of myths, though not all the stories are directly related to each other. Thus Cetus, a constellation that reaches its most prominent position high in the southern sky in November, is connected to the Perseus myth, to Pegasus, and to a number of constellations in the northern sky. Cetus was the sea monster that devastated Ethiopia, the land ruled by Cepheus and Cassiopeia (see pages 28–29).

Unlike Perseus and Cassiopeia, **Cetus** (the Whale) has very few bright stars, but one of them, Mira, is of particular interest. (Mira is a Latin word that means wonderful or astounding.) Mira was the first variable star to be discovered, and it became the prototype of a whole group of variable stars, the Mira-type variables (see pages 150–51). When Mira is at its brightest, it is easy to see with the naked eye; when it is faintest, it can disappear from view.

In the 1960s some astronomers suspected that there might be another wonderful star in Cetus, namely, the star Tau Ceti, which resembles our Sun and is only 12 light-years away. It and Epsilon Eridani were considered the most likely stars to have planets like those of our Sun, planets on which there might be life. American astronomers therefore trained their radio telescopes in the directions of these two stars and listened intently for signs of some distant civilization, but without success.

North of Cetus stands Pisces, another constellation of the zodiac that is traversed by the ecliptic and where planets therefore could turn up. Pisces, too, is an inconspicuous constellation with few stars visible to the unaided eye. In old sky atlases, this constellation is represented by a pair of fish whose tails are tied together by a rope. According to Greek mythology, Aphrodite, the goddess of love, and her son Eros were fleeing from a raging monster but were unable to cross the Euphrates River in Mesopotamia because of high water. Desperate, they

November 15, 10 P.M.
November 30, 9 P.M.

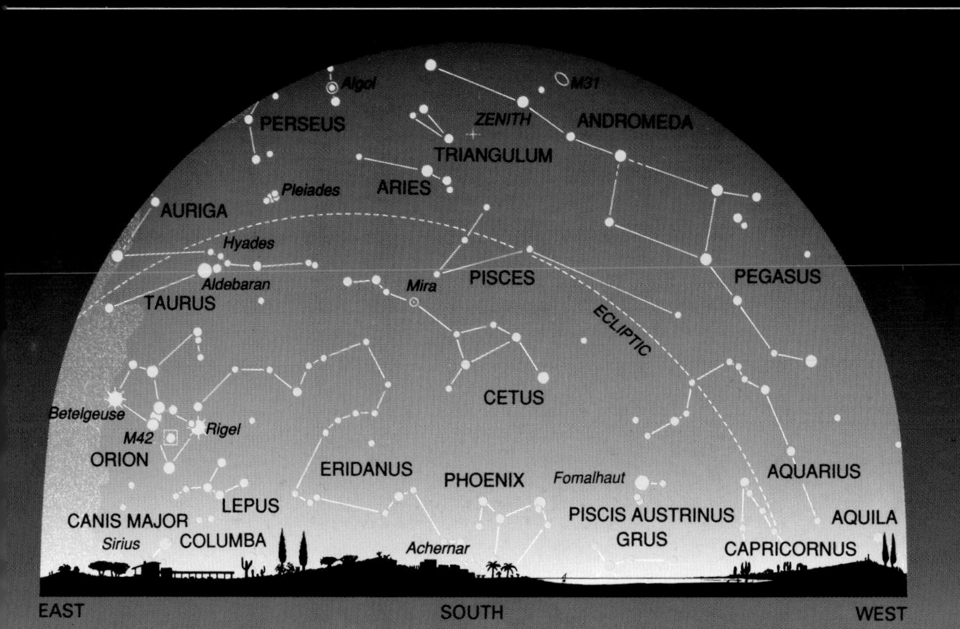

nevertheless plunged into the water, and the sea goddesses took pity on them, sending two fish to their aid. The fish were connected by a rope, which Aphrodite and her son could hold on to. In gratitude for this deed, the gods placed the two fish in the sky. It was in the constellation Pisces that an unusual planetary phenomenon occured in the year 7 B.C., namely, a rare multiple conjunction of Jupiter and Saturn (see page 114). Most modern astronomers now believe that this alignment of the planets in Pisces was the famous star of Bethlehem that announced the birth of Christ. Our present system of counting the years beginning with the birth of Christ—the exact date of which is still the subject of controversy—was not invented until 500 years after the fact.

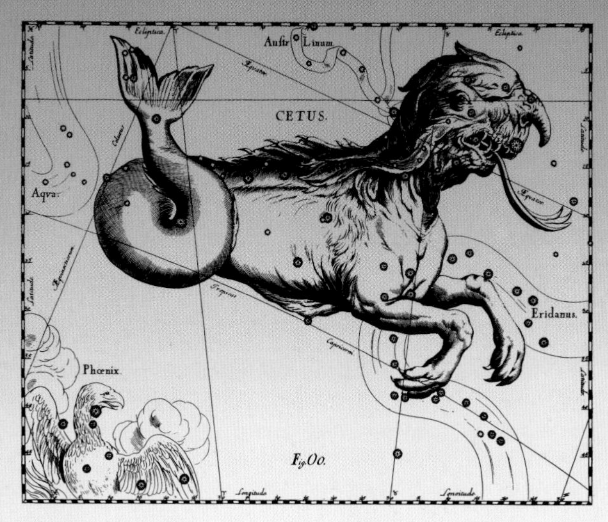

Cetus

December

Star map N II/12

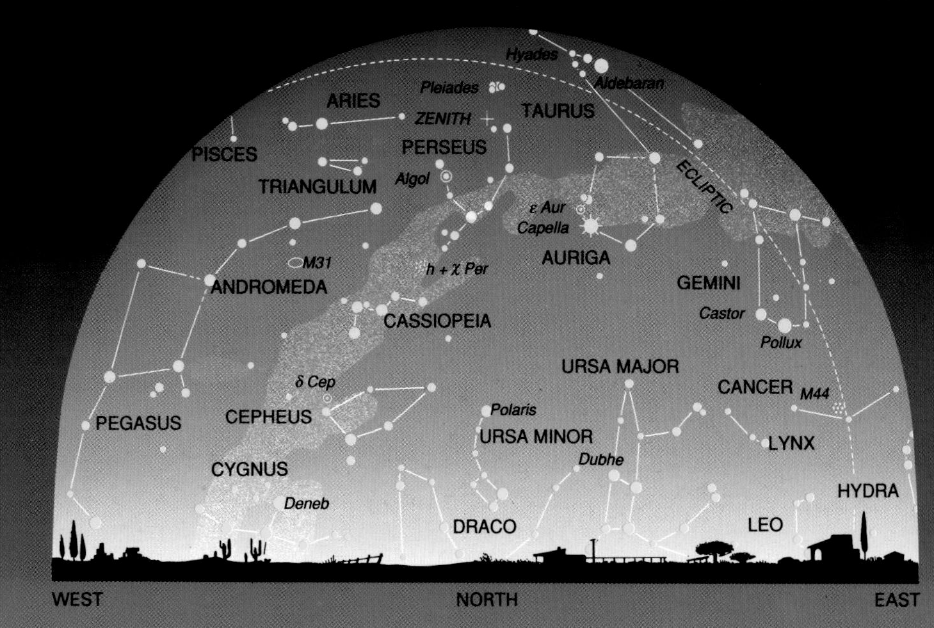

In December, the night sky displays the well-known winter constellations during the early evening hours in an advantageous position in the southeast. Unlike in the fall, the south has again clearly become the dominating region of the sky. The ecliptic arches high across the celestial dome, so that during December the planets are in a good viewing position and the Moon is particularly high and keeps shining a long time over the horizon. December is therefore a good time to watch some of the most splendid constellations of the zodiac, especially Gemini, Taurus, and Aries. **Aries** (the Ram) usually is considered the first of the signs of the zodiac. In earlier times, before the birth of Christ, the Sun was in Aries on the first day of spring, so that that point on the ecliptic, which now usually is referred to as the vernal equinox, was—and sometimes still is—called the first point of Aries. Because of precession, the slow shifting of the celestial sphere's coordinates against the background of the constellations (see pages 104–05), the vernal equinox has meanwhile moved into the constellation Pisces and will in about 400 years reach Aquarius.

The constellation Aries is inseparably linked with the story of Jason and the Argonauts. Their ship, the Argo, accounts for several beautiful constellations in the southern sky, namely Carina (the Keel), Vela (the Sails), and Puppis (the Stern). One of these, Puppis, can be seen just barely over the southeastern horizon (for more on Argo Navis, see pages 82–83). In Boeotia in ancient Greece there once lived a king by the name of Athamas whose realm was visited by a terrible famine. In desperation, Athamas was ready to sacrifice his two children Phrixus and Hella after the children's wicked stepmother had caused messengers to deliver a false oracle to him. But just before the children were about to be killed, the gods sent a ram with a coat the color of pure gold and with the gift of human speech to save them.

December 1, 11 P.M.
December 15, 10 P.M.
December 31, 9 P.M.

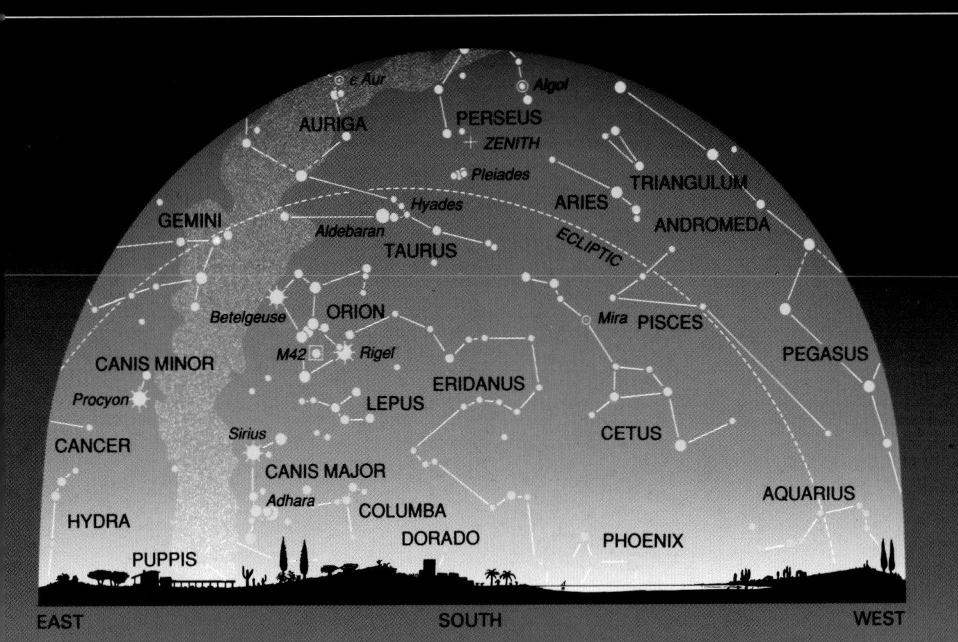

He carried the children aloft and flew eastward with them across the sea. Hella was seized by a sudden panic and lost her hold, plunging down into the water in a spot that to this day is called Hellespont. Phrixus, the boy, was deposited safely in the country of Colchis on the Black Sea. On advice of the gods, Phrixus sacrificed the ram and hung its pelt, the famous golden fleece, in the garden of Ares, the god of war, where a huge dragon guarded it. It was this golden fleece the Argonauts were later commanded by the gods to retrieve, a feat they successfully performed.

Today, Aries is one of the smaller constellations. It has only one relatively bright star, Hamal, which means head of a sheep in Arabic. Hamal is 80 light-years away from us and shines about 70 times brighter than the Sun. The Sun travels past Aries between April 19 and May 13 every year.

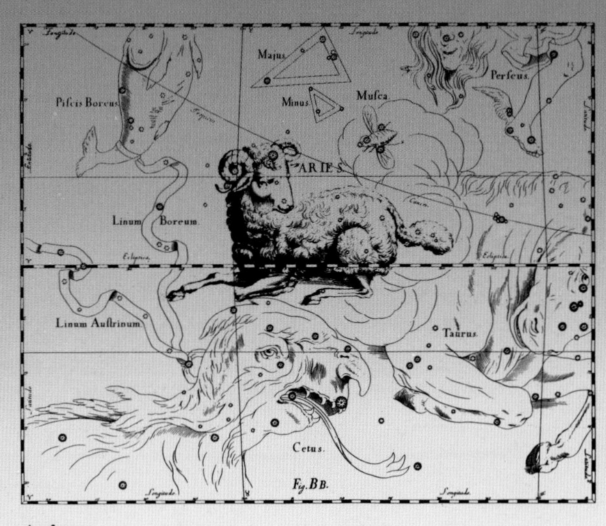

Aries

Star Map Series S

This series of star maps is applicable in the countries that lie between the 10th and the 30th degrees of southern latitude. The night sky of the Southern Hemisphere is quite different from that of the Northern Hemisphere. To the observer, the northern constellations seem to be upside down, as a glance at the maps showing the northerly half of the sky immediately demonstrates. The southern constellations even have different kinds of names from those of the north because they go back to much more recent times, namely the eighteenth century. The abbreviations LMC and SMC stand for the two most obvious celestial objects in the southern sky, the Large Magellanic Cloud and the Small Magellanic Cloud. They can be seen in the sky even at full moon as shimmering patches (see page 137).

The farther south one travels in the S-zone, the higher the stars shown in the maps will appear to be above the southern horizon and the lower they will appear to be above the northern horizon. The charts of series S, like those of the two N series, show the sky as it looks at 11 P.M. at the beginning of the month, at 10 P.M. in the middle of the month, and at 9 P.M. at the end of the month.

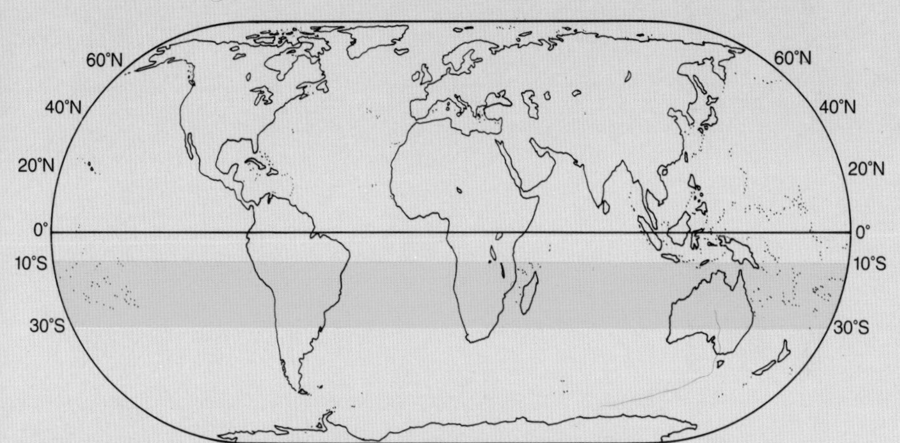

The star maps S apply to the following countries: Peru, Bolivia, Argentina, Chile, South Africa, Zimbabwe, Tanzania, Botswana, Namibia, Madagascar, and Australia.

What the symbols in the star maps stand for:

-●- = Double stars: Pairs of stars that are particularly close to each other.
◉ = Variable stars: Stars that change in brightness.
□ = Nebulas: Clouds of gas and dust that shine in colors.
∴ = Star clusters: Aggregations of many stars concentrated in one place.
O = Galaxies: Star systems (the Milky Way is one of many galaxies).

About the brightness and size of the stars marked in the maps:

0 = ✸ The brightest stars. They are easily visible even in big city skies.

1 = ● 2 = ● 3 = ● 4 = • Scale of brightness 0 to 4. The smaller the number, the brighter the star.

◁ *This wisplike nebula in the constellation Lyra is probably the remains of a supernova.*

January

Star map S 1

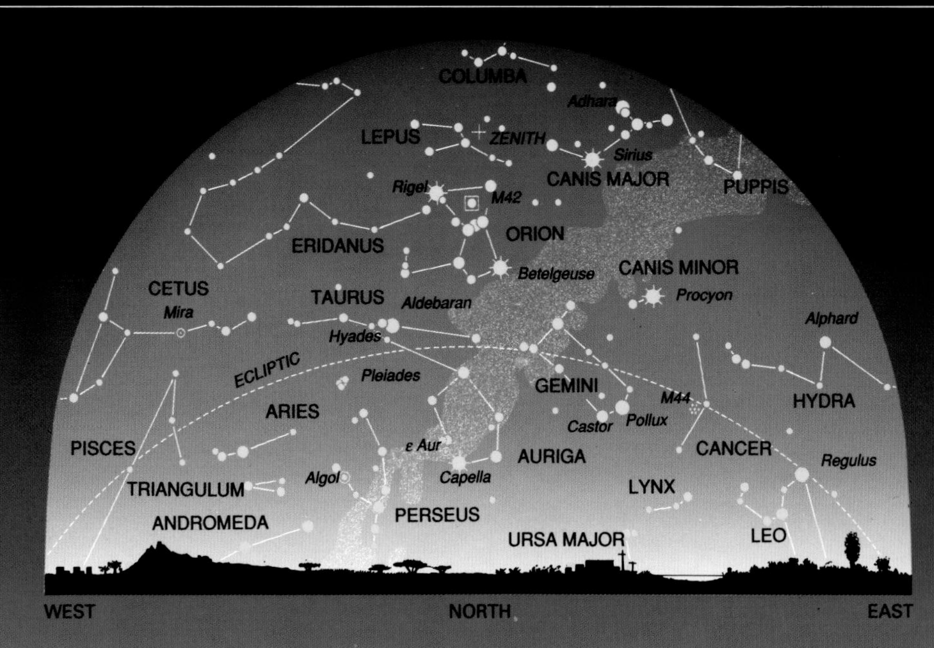

The Sun is closer to Earth in January than at any other time of year. It is only 91 million miles (147.1 million km) away from us. However, the fact that it is now summer in the Southern Hemisphere has little to do with this relative closeness of the Sun, even though the intensity of heat rays does become greater as the distance from their source decreases. But the real reason why there are different seasons of the year is the tilt in Earth's axis (see page 13).

The starry skies of the Southern Hemisphere during the months of January through March are some of the most spectacular anywhere. Seven of the ten brightest stars (all of which are shown in the maps as yellow dots surrounded by small rays) can be seen at an easily viewable height over the horizon. The only ones that are missing are Vega, Arcturus, and Alpha Centauri, though the last mentioned will rise later in the night. The northern part of the sky is dominated by the constellations that include the stars composing what is known in the Northern Hemisphere as the Winter Hexagon. These constellations are Auriga, Taurus, Orion, Canis Major and Canis Minor, and Gemini. In these latitudes and in keeping with the local season, Summer Hexagon would be a more apt term for the Winter Hexagon.

There are many smaller constellations in the southern sky that are named after animals. There are, for instance, **Tucana** (the Toucan), **Phoenix**, **Hydrus** (the Lesser Water Snake), **Volans** (the Flying Fish), **Grus** (the Crane), **Dorado** (the Swordfish), and **Pavo** (the Peacock). These constellations are all grouped around the south celestial pole, which is—in contrast to the north pole—not marked by any bright star but can easily be found with the help of the Southern Cross (see page 22), which is clearly visible at this time of year.

The names of the many exotic animals in the sky of the south do not come from Greek mythology as do the

January 1, 11 P.M.
January 15, 10 P.M.
January 31, 9 P.M.

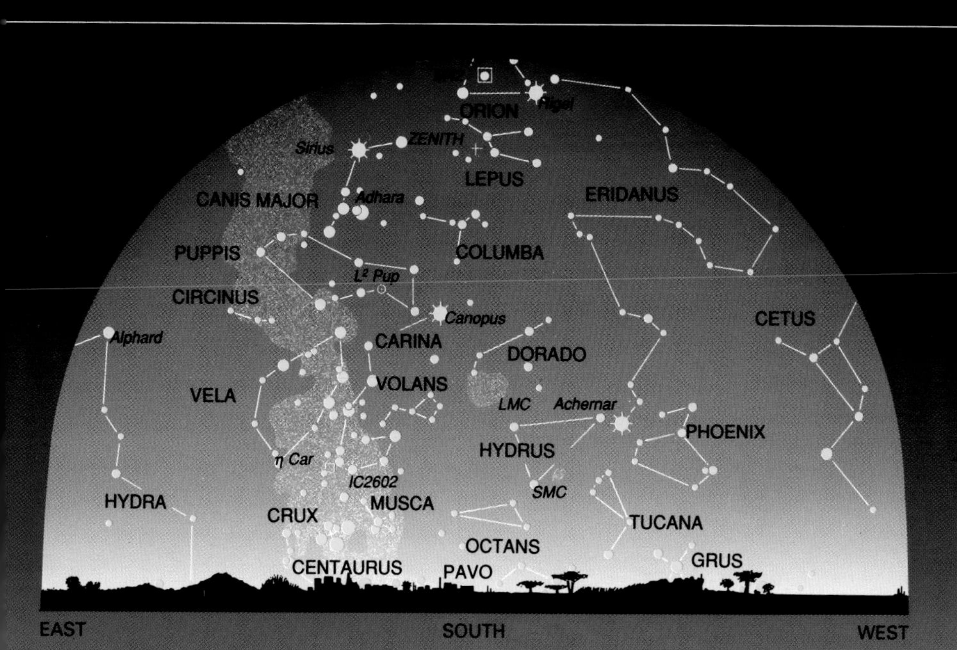

names of constellations in the north, but were invented much later in the seventeenth century. A German astronomer by the name of Johann Bayer, who lived in Augsburg from 1572 to 1625, first presented these constellations to the public in 1603 in his famous star atlas. He used as his sources older reports by Dutch seafarers who had been some of the first Europeans to observe the skies of the Southern Hemisphere in the sixteenth century. Bayer took over various constellations from these accounts, arranging them in rather arbitrary fashion around the south celestial pole, and his draftsmen drew heavily on their imaginations in illustrating them. There is no recognizable resemblance between the star patterns and the animals for which they are named, whereas in the sky of the north the connection between the names and their constellations still can be seen at least to some extent, as in the case of the Ursa Major.

Tucana, Phoenix, Hydrus, Volans, Grus, Dorado, Pavo

February

Star map S 2

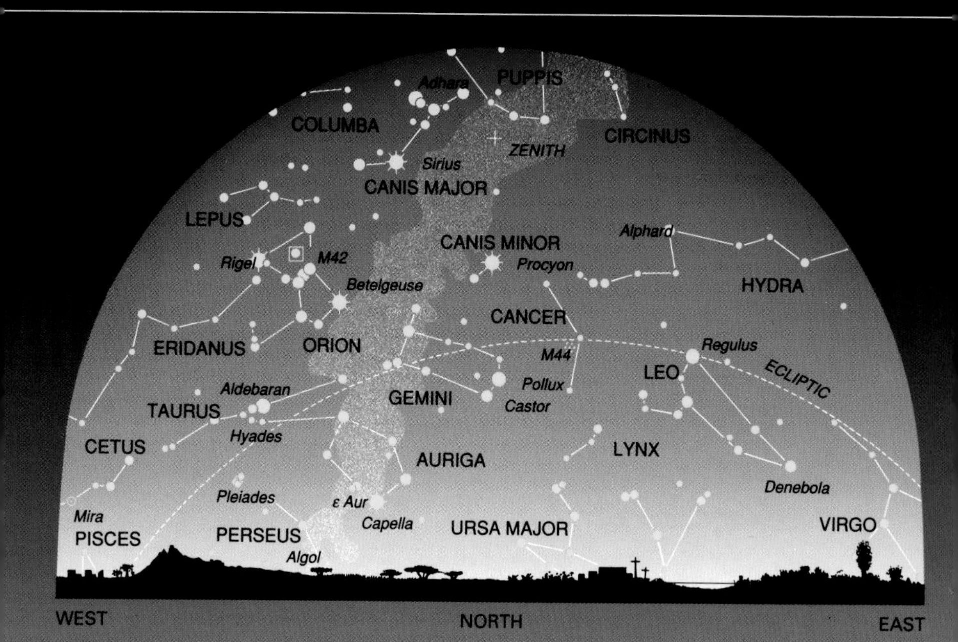

In February, it already is possible to see Centaurus above the horizon in the evening. It, as well as Carina, Vela, and Puppis—parts of Argo Navis—are evidence that some constellations of the southern sky go back to Greek mythology and that not all of them are inventions of the recent past.

Argo Navis (the Ship of the Argonauts) is connected with one of the most exciting adventure stories of Greek mythology. It was originally a single constellation, though today it is split up into four independent parts, which are all adjacent to each other in the sky. A number of other constellations in their vicinity also figure in this famous story.

Jason, the commander of the Argo and its crew, had been sent by his uncle on the dangerous mission of recovering and bringing back to Greece the Golden Fleece. Only after successful completion of this task was he to be allowed to rule over his father's realm. The Golden Fleece was the pelt of the ram that had been sacrificed to the gods after an adventuresome flight across the seas (see pages 76–77). The fleece was in Colchis, fiercely guarded by a dragon.

Jason, aided by Athena, had a big, powerful ship built—the Argo—and advertised throughout Greece the adventure he was setting out on. The most distinguished heroes of the country, including, among many others, Hercules, Castor, and Pollux (see pages 58–59), responded to the call and joined the Argo's crew. Eventually, of course, Jason succeeded in his mission after killing the dreadful dragon. But before arriving in Colchis, the Argonauts had to pass through many narrow straits where their ship was in great danger of being dashed against the cliffs. However, Athena sent a dove to point the way and guide the Argonauts safely past the rocks, and this dove now stands in the sky, too, near Argo Navis, in the form of the constellation Columba.

The most interesting of the four constellations making up the ship Argo is Carina (the Keel). In it shines

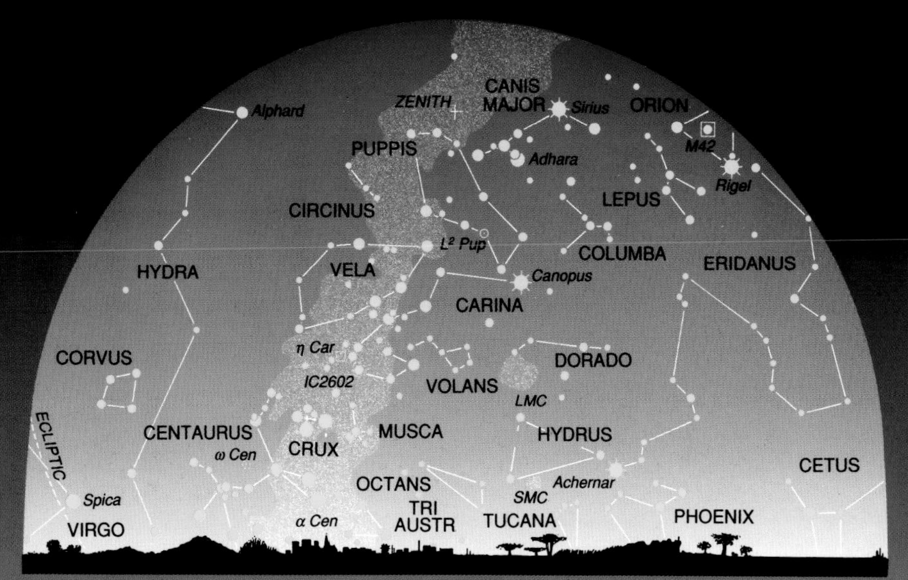

the second brightest star of the heavens, Canopus (see page 133). Another interesting object in this constellation is a large and brilliant star cluster, technically referred to as IC 2602 (this means that it is the 2,602nd entry in the Index Catalogue, a comprehensive listing of star clusters). IC 2602 often is called the southern Pleiades in allusion to the famous Pleiades visible in the sky of the Northern Hemisphere (see pages 147 and 145). It is best to look at IC 2602 through binoculars. Close to the impressive star cluster there is a huge nebula that surrounds the star Eta Carinae. Eta Carinae is an object of the Milky Way, but there is at this point no definite knowledge about its nature (see page 142).

Argo Navis

In February 1987, a star flared up in the Large Magellanic Cloud (LMC on the map above). This was the first bright supernova to appear in Earth's skies in almost 400 years, giving modern astronomers an opportunity to study this kind of object with today's sophisticated instruments.

March

Star map S 3

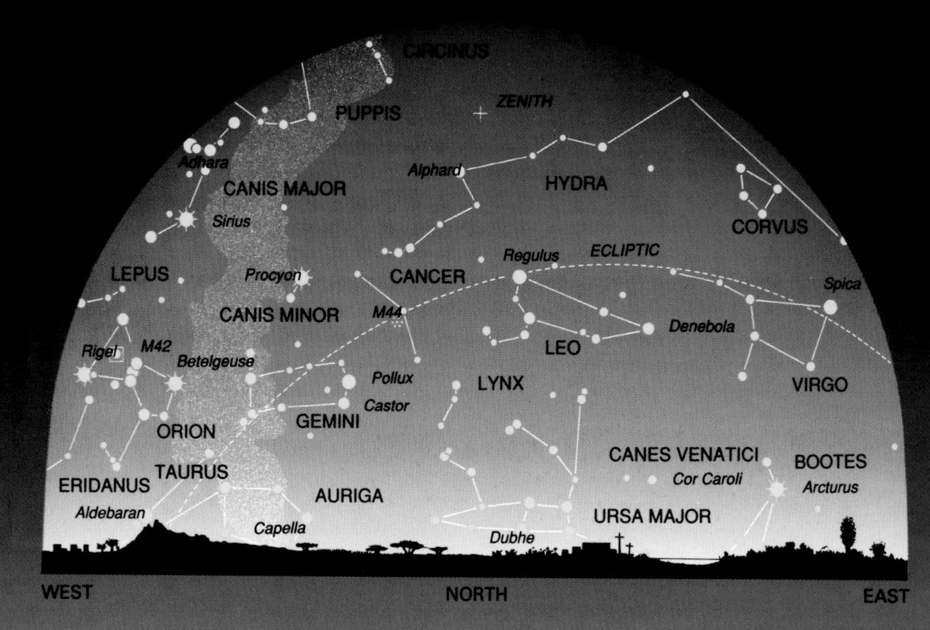

In March, the constellations making up Argo Navis dominate the southern sky. **Columba**, which pointed the way through the straits when the Argonauts sailed eastward to retrieve the Golden Fleece, is in the southwest. Above it, Puppis rises up between the two brightest stars of the sky, Sirius and Canopus. Puppis (which means poop or stern in Latin) contains an unusual star, one of the few variables in which one can observe the changes in magnitude with the unaided eye. This is L2 Puppis.

L2 Puppis is a Mira-type variable. Its behavior is similar to that of the wonderful star Mira in Cetus (see pages 140–41). L2 Puppis fluctuates in brightness in a regular cycle of approximately 140 days. At its maximum brightness it is easy to see, but in its dimmest phase its light is at the borderline of what our eyes can detect.

In the southeast, Centaurus and the Southern Cross (Crux) slowly approach their highest point in the sky. The Southern Cross is found in one of the starriest regions of the Milky Way. But within the area of the Southern Cross there is also a particularly dark spot, the so-called Coalsack. There, huge dark clouds of dust block the light of the stars behind them. An especially impressive view of the Coalsack Nebula is obtained through binoculars because one almost has the sense of being in a hole within the band of the Milky Way. The Coalsack Nebula spills into the constellation Musca (the Fly), a rather unusual name first introduced in 1624 by Jakob Bartsch, a son-in-law of Germany's most renowned astronomer Johannes Kepler. The Southern Cross owes its fame primarily to its usefulness as a navigational aid on the seas because its axis points toward the south celestial pole and therefore indicates where south is on the horizon (see pages 22–23). Apart from this, the Southern Cross is not quite as impressive as those in the Northern Hemisphere often imagine. Although the Southern Cross is considered a symbol of Christianity, it already was known to

March 1, 11 P.M.
March 15, 10 P.M.
March 31, 9 P.M.

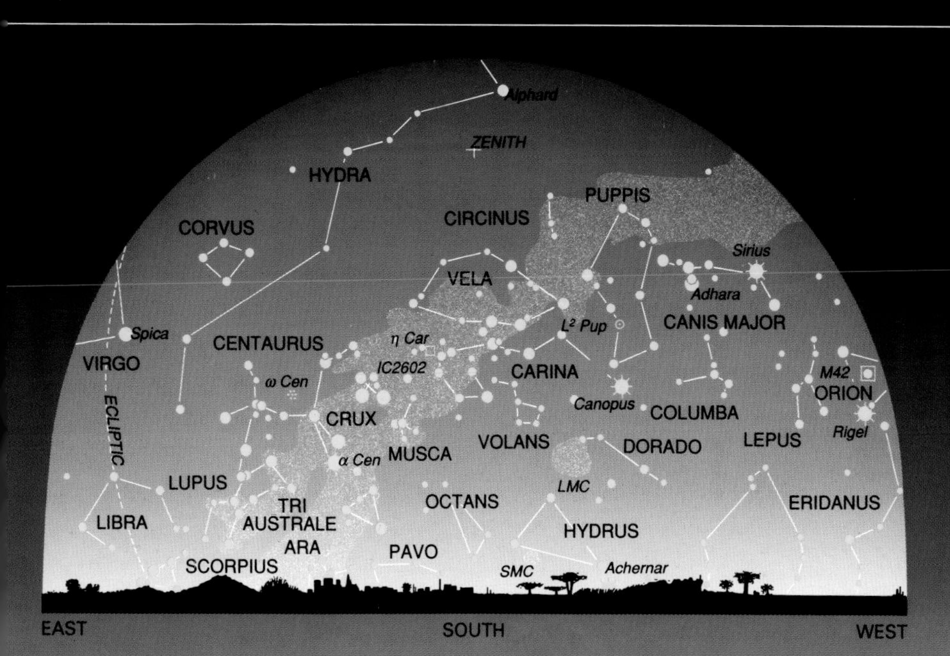

the astronomers of antiquity. But they regarded it as part of Centaurus. Today the Southern Cross holds the record of being the smallest of the 88 constellations in the sky (Hydra is the largest).

In 1942, an exceptionally bright nova appeared in Puppis, a star that blazed to 15 million times its previous brightness within a few days. This amazing outburst of luminosity led astronomers to wonder whether they had witnessed a supernova (see pages 142 and 149), but according to the current state of knowledge this was not the case. At this point all that remains of Nova Puppis 1942 is an extremely faint star surrounded by some bright, thin clouds of gas that were ejected during the outburst.

Columba

April

Star map S 4

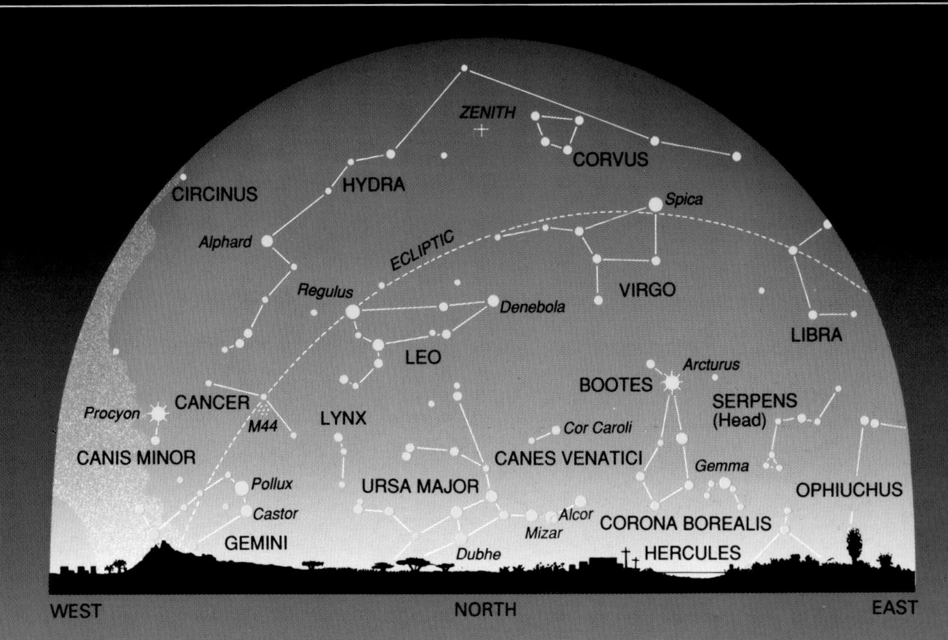

In April, a glance into the sky straight overhead reveals Hydra in its full, huge length. Hydra is the largest constellation, but it is made up of relatively dim stars (see pages 62–63). Toward the south, **Centaurus** dominates the sky.

Centaurs were fabled beasts from Greek mythology. Half man and half horse, most of them were of gross and violent nature and responsible for much trouble. But there was one exception, the centaur Chiron. He was friendlier and gentler than the rest of his kind, in possession of great wisdom, and a teacher of the art of healing. But his fate took a cruel turn. One day he encountered Hercules, who accidentally dropped one of his poisoned arrows on Chiron's foot. Although Chiron was a master physician, he was unable to heal this wound. His suffering became so unbearable that he wished for death even though he was of divine birth and therefore immortal. By choosing death, he freed Prometheus from his torment.

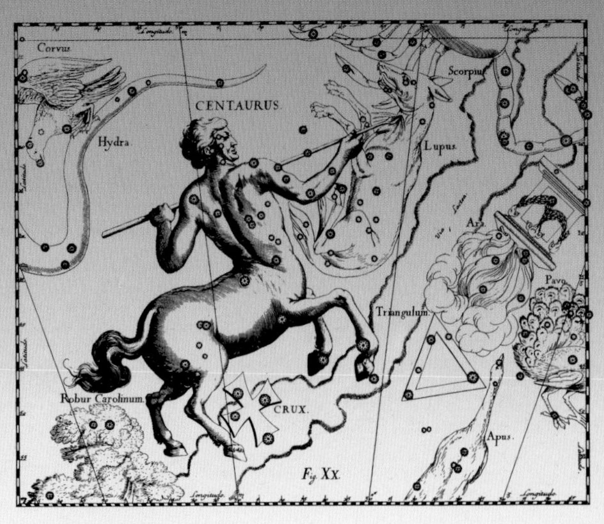

Centaurus

April 1, 11 P.M.
April 15, 10 P.M.
April 30, 9 P.M.

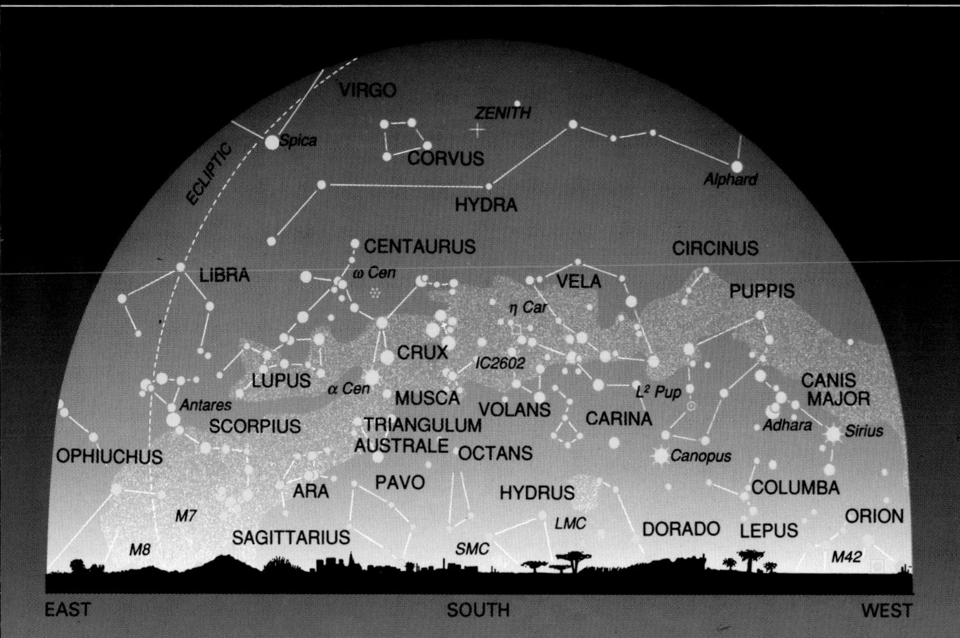

Prometheus had been condemned to eternal suffering because he had stolen fire from the gods and given it to mankind. As a reward for this selfless act, the gods transformed Chiron into the constellation Centaurus.

The constellation Centaurus is famous today primarily for its most brilliant star, Alpha Centauri, one of the ten brightest of the sky. It appears so bright in part because it is closer to our solar system than any other star (see pages 130–31). Centaurus also contains another unusual celestial object, Omega Centauri, the brightest globular star cluster in the sky. The designation Omega (a Greek letter) shows that this star cluster originally was taken to be a star (see pages 21 and 147), and to the unaided eye it appears to be a star. However, by using a pair of ordinary binoculars one can see its more complex structure, though it appears fuzzy.

Globular clusters are exceptionally compact groupings of huge numbers of stars. This is in contrast to the less dense open star clusters (see page 147), such as IC 2602. The latter appears at this time of year a little farther to the west in Carina. Omega Centauri, which is one of the closest globular star clusters, is about 15,000 light-years away. Intensive observation has shown that Omega Centauri consists of at least 100,000 stars. This does not mean that all the stars in the cluster are packed closely together. Even in the interior of a globular cluster, individual stars are separated from each other by many billion miles. Still, the celestial view inside a globular star cluster would have to be extraordinary because the sky would be full of stars comparable to Sirius and Canopus and even brighter ones.

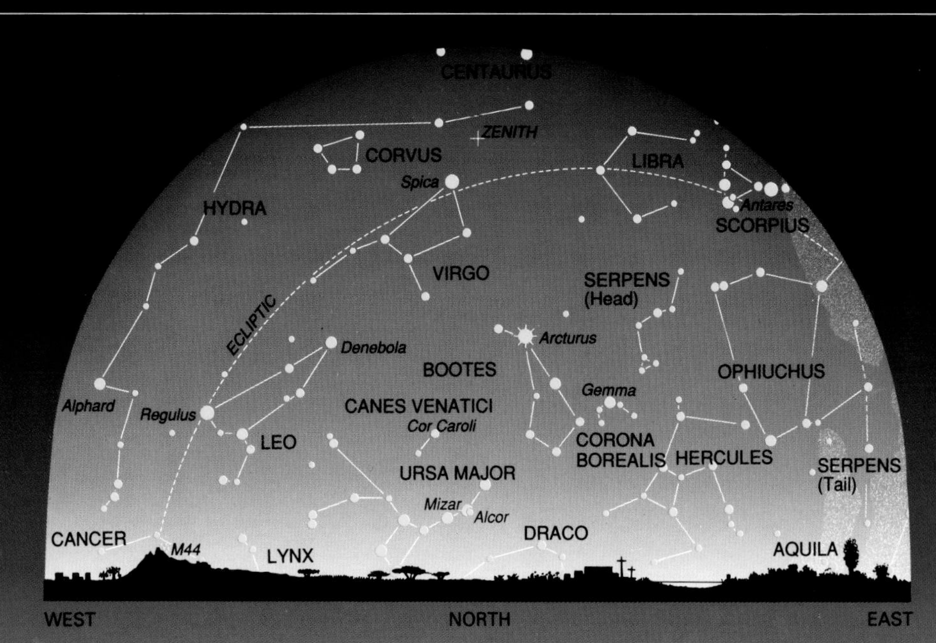

Argo Navis with its three constituent constellations Carina, Puppis, and Vela already is quite low in the western sky during this month. Only Vela still reaches up a little higher. West of Vela is the obscure constellation Circinus (the Compass). In spite of its proximity to Argo Navis and the fact that compasses are associated with ships, there is no direct connection between these constellations because at the time when the Argo sailed the seas there was no such thing as a compass.

The name Circinus and that of Octans, the constellation that lies due south, are recent inventions. We owe them to a French astronomer who studied the sky of the Southern Hemisphere intensively from the Cape of Good Hope between 1751 and 1753 and invented the last of the 88 names still used today. This was Nicolas Louis de Lacaille (1713–1762), who made up a total of 14 new constellations. When de Lacaille first watched the southern sky there were only a few constellations in it to look for, namely those that Johann Bayer had created on the basis of descriptions by Dutch navigators (see page 81). De Lacaille was blessed with a lively imagination. He wrote in his catalog of stars: "To fill in the large empty spaces between the old constellations I introduced new ones. The figures I used to represent them are mostly instruments used by practitioners of art and science." Thus he placed in the heavens the Furnace (Fornax), the Clock (Horlogium), the Compass (Circinus), the Octant (Octans), the Microscope (Microscopus), and even an Air Pump (Antlia). These constellations still are recognized today and—compared to the more imaginative names of the northern constellations—give the southern sky a rather prosaic quality, especially since all of de Lacaille's constellations are made up of such dim stars that they can be made out only with great effort. Only two of de Lacaille's 14 constellations have been entered in these maps, namely Circinus and Octans, the latter

May 1, 11 P.M.
May 15, 10 P.M.
May 31, 9 P.M.

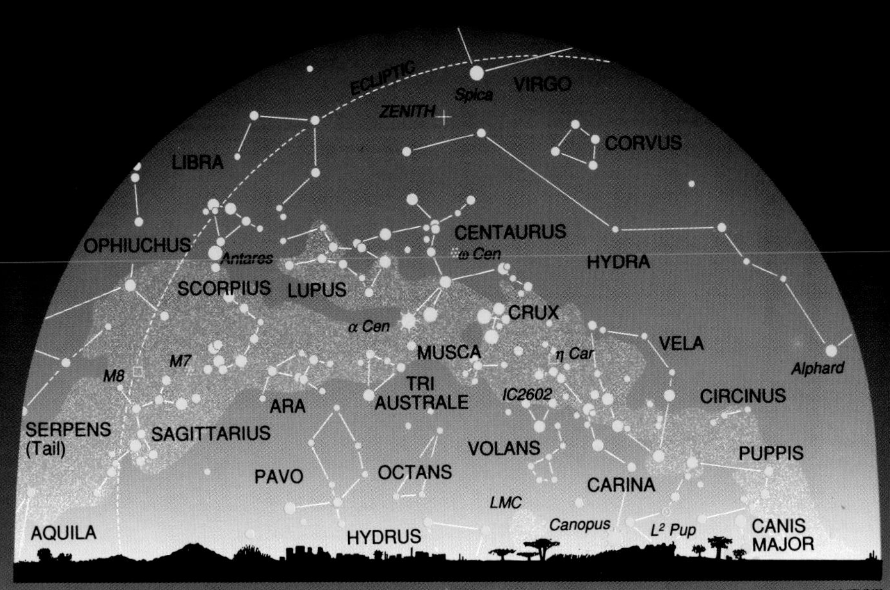

an instrument for determining altitudes in the sky by using angles of 45 degrees, or an eighth of a full circle (octo is Latin for eight). Better known than Octans is **Sextans**, which was named by Hevelius. The sextant, an instrument similar to the octant, uses 60-degree angles (a sixth of a full circle).

Neither Circinus nor Sextans is remarkable for anything in particular, and Octans, too, displays only a few faint stars. Its claim to fame rests on the fact that the south celestial pole falls within its confines. However, Octans is of no use when trying to find the south celestial pole quickly. For this purpose, one turns to the Southern Cross and extends the line of its axis by four times its length (see page 22).

In the northern part of the sky, Arcturus is the most prominent spot of light during May. Together with Spica in Virgo and Regulus in Leo, it forms the Vernal Triangle. In the Southern Hemisphere it would, of course, be more logical to call it the Autumnal Triangle.

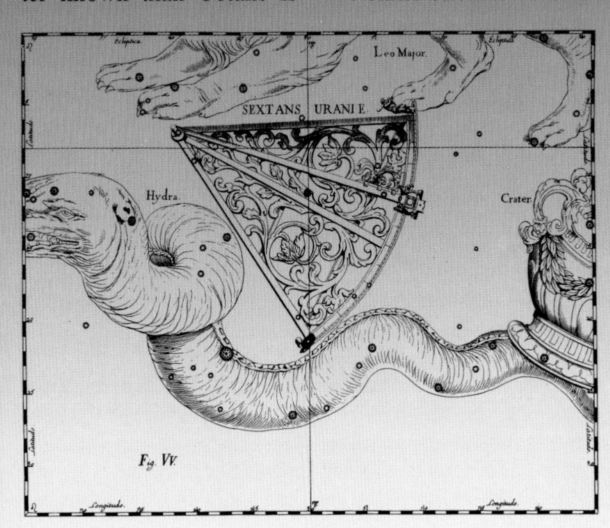

Sextans

June

Star map S 6

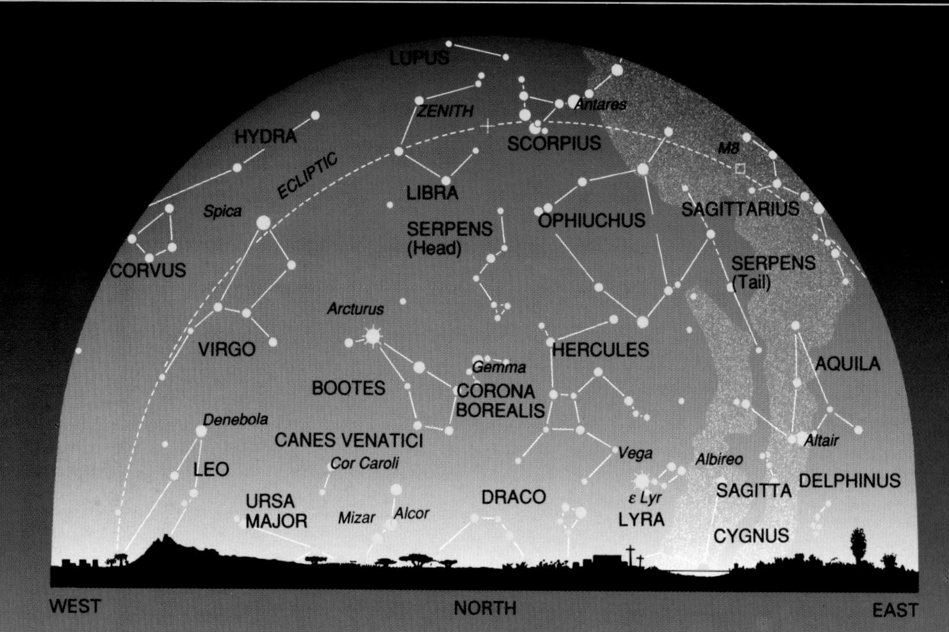

At the beginning of winter in the Southern Hemisphere, the ecliptic rises up steeply and passes through the zenith. This means that the planets that are visible in June are in an exceptionally favorable position, almost directly overhead.

The dominant constellations in the south are now Centaurus, the Southern Cross, and—almost at the zenith—Scorpius and Sagittarius. The Milky Way runs through all these constellations, presenting an impressive aspect in June. The most conspicuous objects in the Milky Way, the globular star cluster Omega Centauri and the open star cluster IC 2602 in Carina, as well as the gigantic nebula M 8 in Sagittarius and the variable star Eta Carinae are all in positions where they can be seen clearly through binoculars. Altogether, the southern part of the sky is filled in June with a variety of objects that invite closer inspection. In addition there are **Triangulum Australe**, (the Southern Triangle) and **Pavo**, (the Peacock), constellations first described in 1603 by the German astronomer Johann Bayer, who based his work on accounts of early seafarers.

There is a rather interesting story concerning the Southern Triangle. According to it the Dutch writer Peter Caesius suggested in the seventeenth century naming these three stars after the Old Testament patriarchs Abraham, Isaac, and Jacob. The stars forming the Southern Triangle therefore also are known as the Patriarch Stars. As this story shows, the constellations have not always been as immutably defined and named as they are today. There were not only many attempts to create new constellations, as Johann Bayer did in the case of Triangulum Australe and Pavo, and, later, Johannes Hevelius—not to mention the Frenchman de Lacaille and his scientific instruments—but also suggestions to abolish or rename old ones.

In the seventeenth century, there was a movement to replace old constellation names that were deemed heathen with Christian ones.

June 1, 11 P.M.
June 15, 10 P.M.
June 30, 9 P.M.

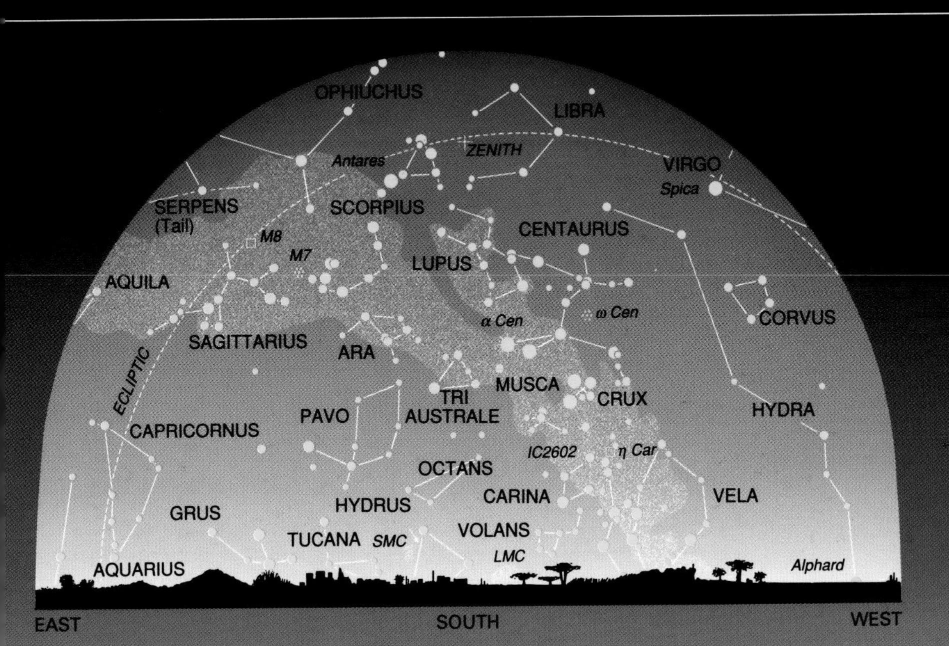

The 12 constellations of the Zodiac, for instance, were to be called after the twelve apostles. In 1627, the German astronomer Julius Schiller published a work called *The Christian Night Sky*. According to this book, Argo Navis, whose sails (Vela) are the last part of the ship to sink below the horizon during the early evening hours in June, was to be turned into Noah's Ark. Further, Schiller fashioned the archangel St. Raphael out of the constellations Tucana and Hydrus and the Small Magellanic Cloud. Lyra, which appears in June, became Christ's manger. However, Schiller's attempt to Christianize the starry sky did not prove very successful.

Ara (the Altar) is one of the most southerly constellations that were known in the ancient world. It can just barely be seen from the Mediterranean region. Ara was said to have been the place where the Olympian gods formed an alliance to defeat the Titans. In commemoration of their victory, the gods placed the altar in the heavens.

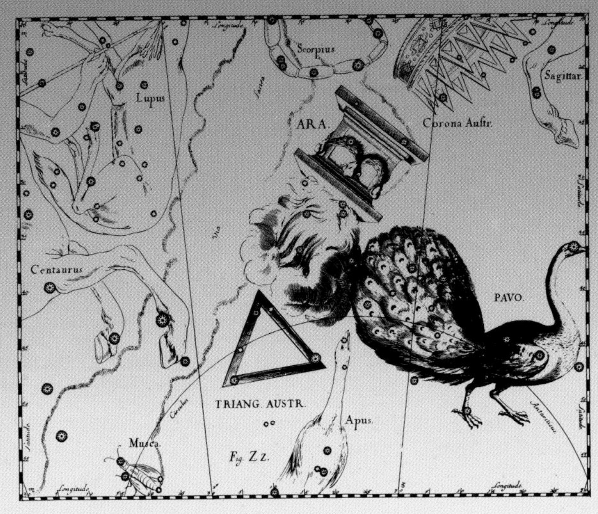

Triangulum Australe and Pavo

July

Star map S 7

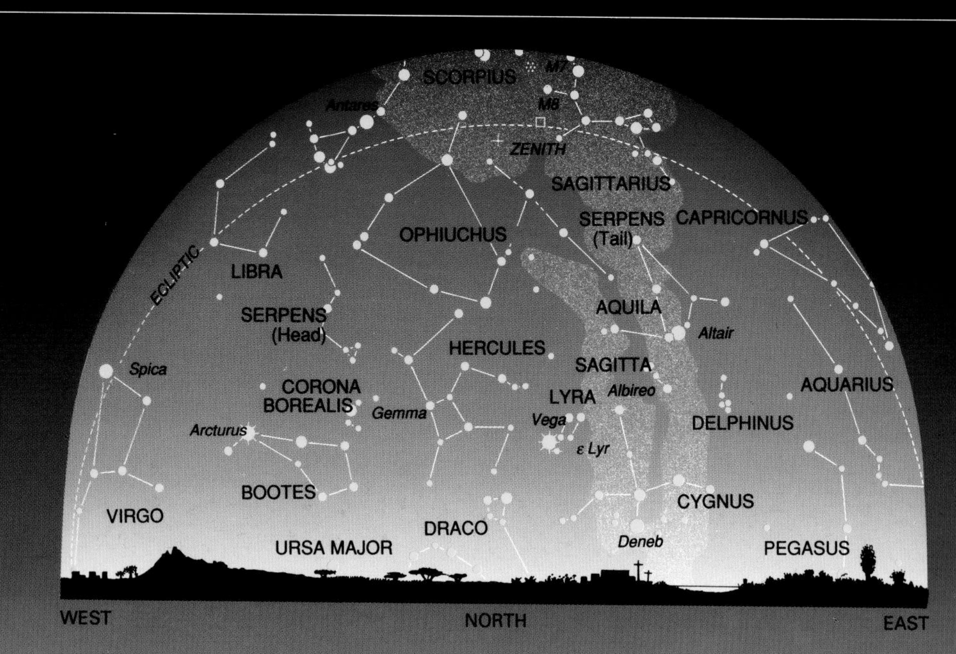

Starting in July, the view into the sky directly overhead is particularly rewarding. The ecliptic runs through the zenith, where there are also two famous constellations of the zodiac with a number of unusual celestial objects. They are Scorpius and, of special interest, **Sagittarius** (the Archer). Pushing between these two constellations and also within the belt of the zodiac, is Ophiuchus, which does not, however, add any bright stars to the ecliptic the way Scorpius does. As part of the redrawing of all the constellations' boundaries in 1930, Ophiuchus was assigned a space between Scorpius and Sagittarius and thus became the thirteenth constellation in the zodiac.

Sagittarius is one of the most interesting constellations of the zodiac, for it lies against the center of the Milky Way. July is a particularly good time to observe the Milky Way anyway because its faintly luminous band also crosses the zenith—as does the ecliptic—rising from Cygnus in the northeast, passing through Sagittarius and Scorpius, and descending down to Carina in the southwest. The Milky Way is a vast star system or galaxy, 100,000 light-years in diameter, on the outer regions of which Earth and our Sun are located (see page 140). The center of the galaxy is about 28,000 light-years away from our solar system, but it cannot be seen because the view is blocked by too many stars as well as by gas and dust nebulas.

Consequently it is not known at this point exactly what lies hidden in the center of the Milky Way. It is possible that some explosion-like processes took place there that produced huge masses of gases. This theory has been deduced in part from the analysis of radio waves that, unlike visible light, can penetrate directly to us from the center of the galaxy. Astronomers estimate that the core of the Milky Way contains a mass equivalent to that of 500 million suns.

Sagittarius, in which lie a number of the most beautiful celestial objects—the gas

July 1, 11 P.M.
July 15, 10 P.M.
July 31, 9 P.M.

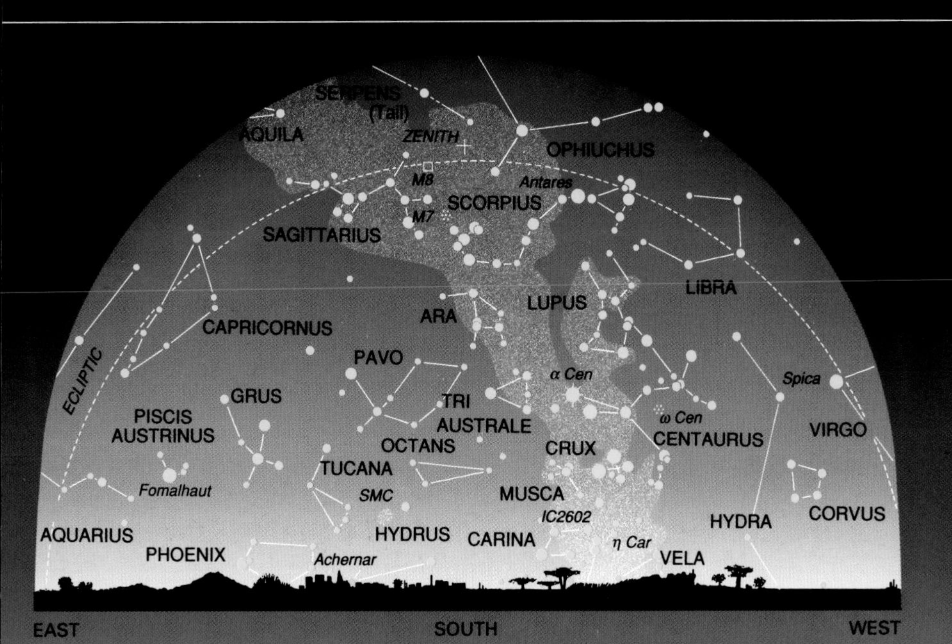

nebula M 8 (see page 142) and the star cluster M 7 (see pages 148 and 21), for example—is one of the classical constellations of the zodiac. Sagittarius usually is depicted in the shape of a centaur—a half human, half horse. In this respect it resembles Centaurus, which is still fully visible in July in the southwestern sky.

The stories associated with Sagittarius, centering around the wise centaur Chiron, are the same ones that are told in connection with the constellation Centaurus (see pages 86–87). The Chinese, on the other hand, saw a tiger, which is a figure of their zodiac, in place of Sagittarius, and Julius Schiller, in his abortive attempt to Christianize the starry sky, replaced Sagittarius with the apostle Matthew. Two thousand years before our era, the Babylonians saw in the constellation we call Sagittarius the King of War, Nergal, which they represented in the shape of a centaur. The conception of Sagittarius as a fabled creature therefore must be of great antiquity.

August

Star map S 8

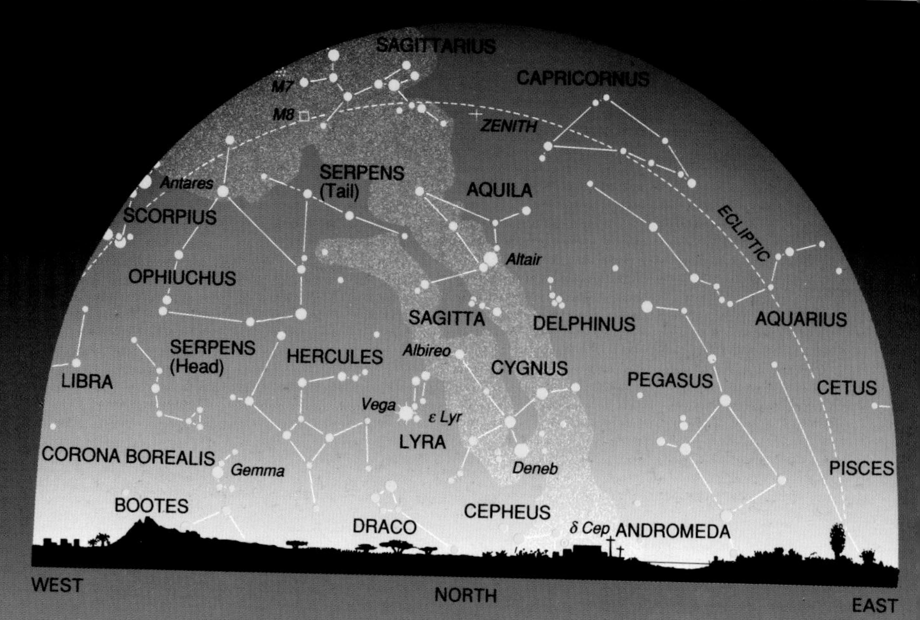

August is the month of the Perseids, a famous meteor shower (see page 139). Between August 10 and 12 a large number of meteors or shooting stars seem to emanate from the constellation Perseus. Unfortunately Perseus is not visible in August in the early evening. It doesn't rise until about 2 A.M, when it appears in the northeast (see table, page 20).

The Milky Way still presents a glorious view in the southern part of the sky. Sagittarius and Scorpius, where the Milky Way shows its greatest variety, are still almost directly overhead. In the southeast, Achernar, the brightest star in the constellation Eridanus, is again plainly visible. Above it, one can spot two other constellations that appear quite prominent because they are in an otherwise starless area of the sky. They are **Piscis Austrinus** (the Southern Fish) and **Grus** (the Crane).

The Crane, named by Dutch sailors, is one of the animal constellations first described in 1603 by Johann Bayer. Piscis Austrinus, on the other hand, has been known as a constellation since ancient times, but its origin is no longer very clear. In older representations, as in Hevelius's star atlas, the Southern Fish usually is shown with its mouth wide open, drinking the water poured into it by neighboring Aquarius. Fomalhaut, whose name means the fish's mouth, is the brightest star in Piscis Austrinus and, in fact, in a large area of the surrounding sky. It is not a particularly large star and is separated from our solar system by only 22 light-years. It is estimated to be about twice the size of our Sun and about 14 times as luminous. In 1983, a satellite launched above Earth's atmosphere to conduct an infrared search of the sky detected a dust nebula around Fomalhaut. This discovery suggests that a planetary system similar to our own may be in the process of forming around Fomalhaut. On such planets that orbit around and derive their energy from a star, conditions might exist that could give rise to some kind of life

August 1, 11 P.M.
August 15, 10 P.M.
August 31, 9 P.M.

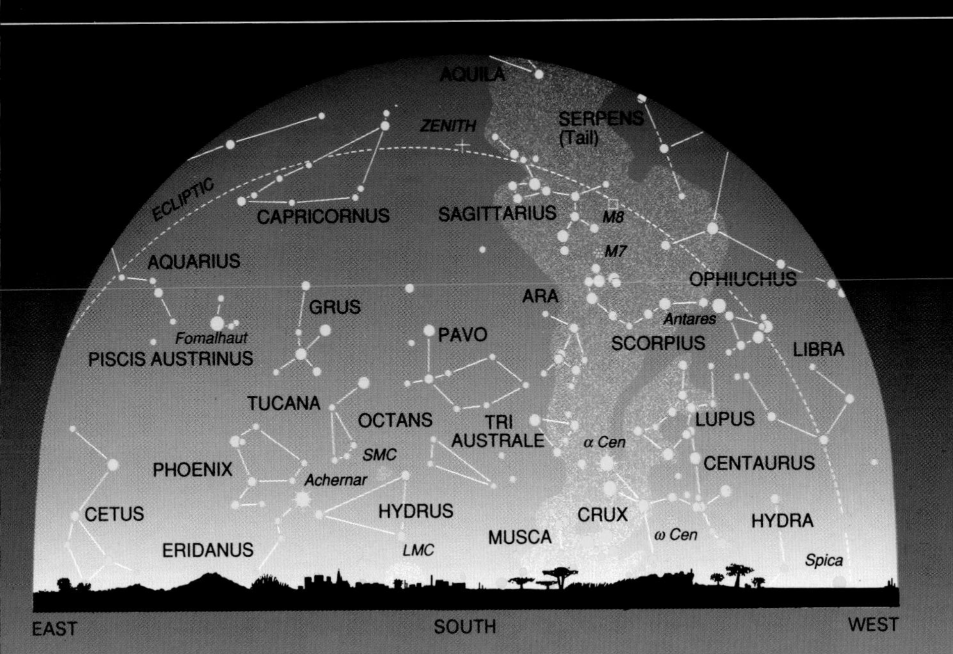

forms. Fomalhaut, together with a star in Eridanus (see pages 98–99) and another one in Cetus (see pages 74–75) are therefore among the star systems in which life might conceivably exist. However, at this point in time we have no sign indicating that this is the case. An expedition to Fomalhaut—or to other stars that may be forming planets—is out of the question. Even if we had space vehicles that traveled at speeds approaching that of light, the journey would take 22 years.

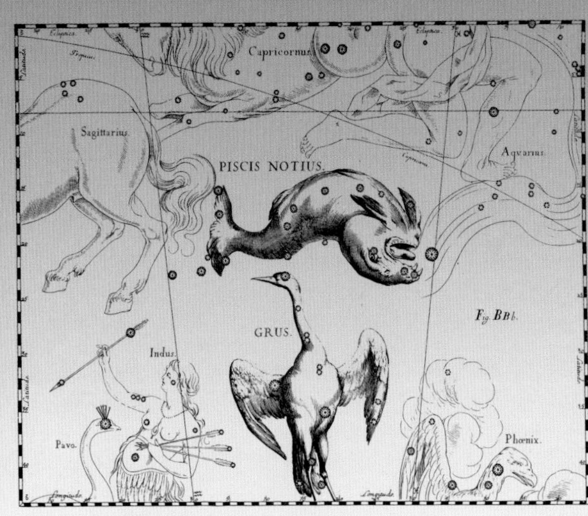

The constellation Grus has little of special interest to offer. But its stars are relatively bright, so that this constellation is easy to pick out in this otherwise starless region of the sky. The area around Grus thus presents a clear contrast to the sky around Sagittarius and Scorpius, through which runs one of the brightest sections of the Milky Way.

September

Star map S 9

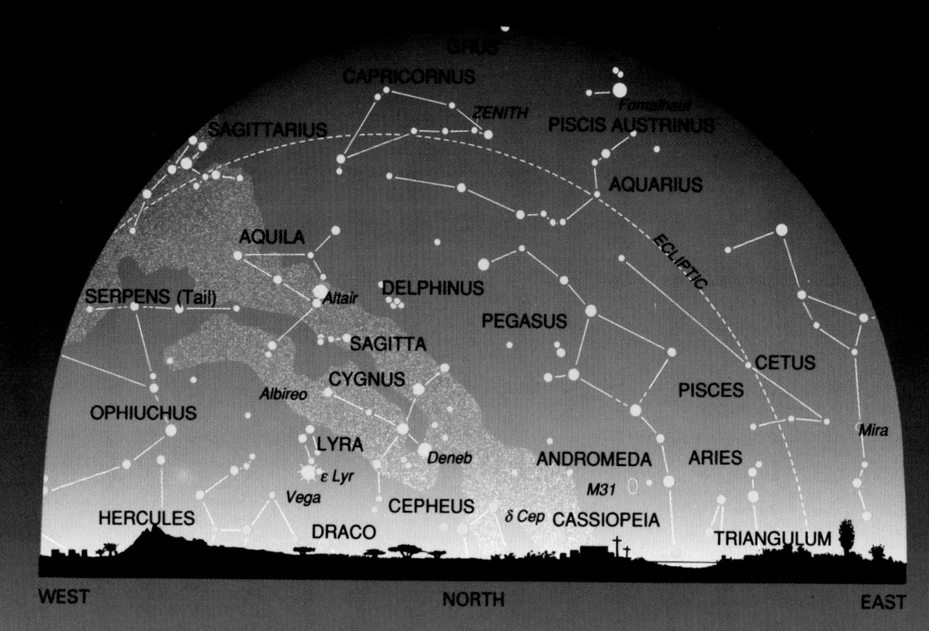

September, which ushers in autumn in the Northern Hemisphere and marks the beginning of spring on the southern half of Earth, is also the month of the New Year for many cultures. Many calendars other than the one we go by begin the new year at the onset of fall or spring. The Jewish and the Persian calendars are examples.

By September the brightest stars in the southern part of the sky have moved on and in the north the brightest ones have not yet returned. Unlike in January or March, very few of the brightest stars of the southern sky are visible. They are Alpha Centauri, which can be seen shortly before it sets in the southwest, Achernar in the southeast, and Vega in the north. A constellation that is not among the most spectacular but has an interesting history, now can be seen low in the southwest. It is **Lupus** (the Wolf).

The name Lupus is of relatively recent origin. Among the Greeks and Romans, this constellation was an unspecified wild animal. Perhaps it was the wolf—or rather she-wolf—that saved Romulus and Remus, the founders of the city of Rome from dying of hunger after they had been abandoned as babies in the forest. This wolf later became an important symbol of the Roman Empire. When an attempt was made much later to bring the sky under the sphere of influence of the Christian faith, Lupus became Benjamin, the ancestor of one of the 12 tribes of Israel.

Almost 1,000 years ago, in the year 1006, Lupus was the site of one of the biggest supernovas ever to be observed. Many who witnessed it in Arabia and in China were so struck by the event that they left precise descriptions. The Arabian astronomer Ali ibn Ridwan of Cairo, for instance, wrote: "At the beginning of my education I witnessed a unique spectacle. A huge star appeared in Scorpius, in the opposite direction of the Sun. This star was round in shape and about two or three times the size of Venus. It lit up the horizon, and its light flickered strongly. It

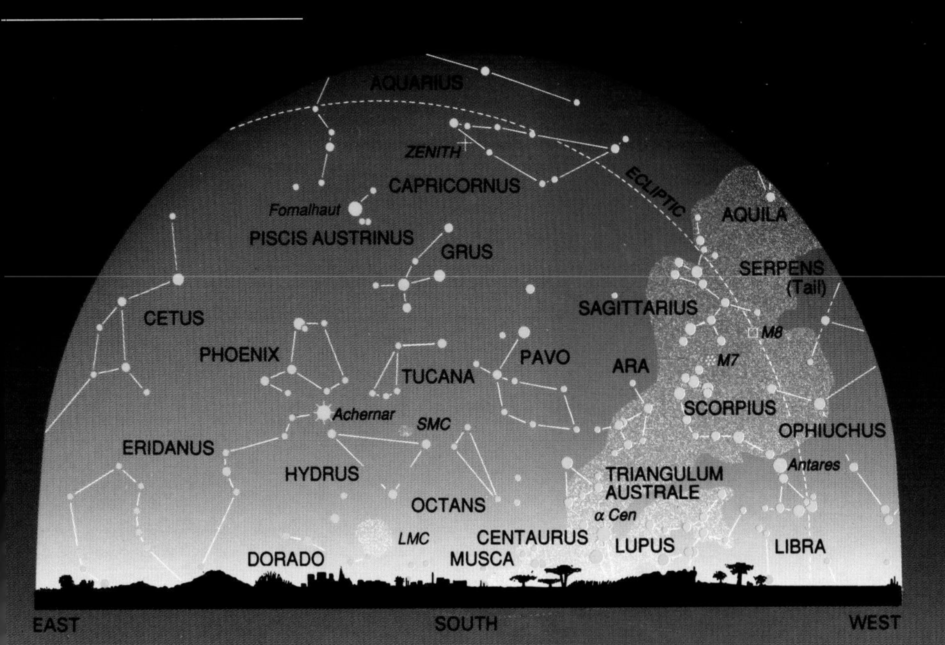

was more than a quarter of the Moon's brightness."

Supernovas are the most spectacular events involving fixed stars. What happens is that a star explodes suddenly and unexpectedly, throwing off into space most of its matter, a process that briefly releases unimaginably vast quantities of energy. In the course of recorded history, only very few supernova explosions visible to the naked eye have been observed. The most recent one took place in 1987 in the Large Magellanic Cloud (see page 137). The last supernova to have been seen in our galaxy occurred in 1604 in the constellation Ophiuchus (see pages 66–67).

Astronomers have been trying for a long time to track down remains of the supernova of 1006 in the constellation Lupus, but so far no conclusive evidence has been found because the position given in the old reports is too imprecise.

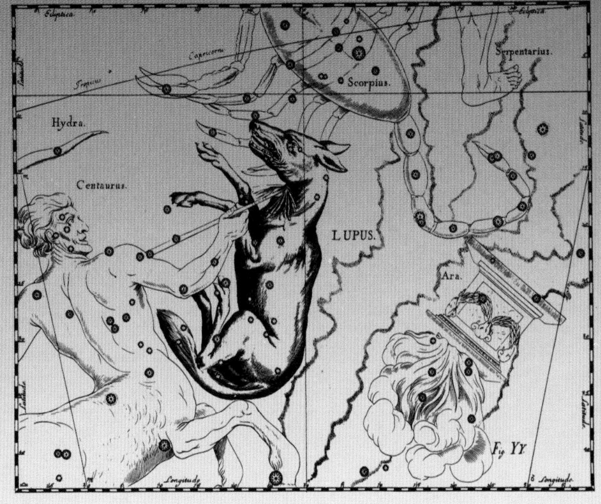

Lupus

October

Star map S 10

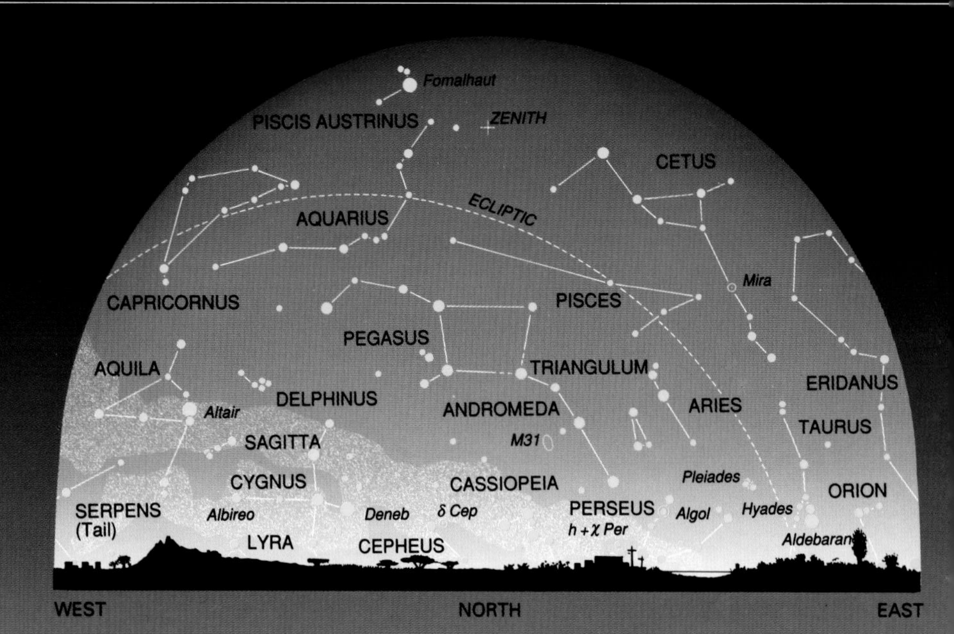

The constellation Eridanus (the River) first can be seen in its full length in the southeastern sky in October. It is not the largest constellation of the sky (this rank belongs to Hydra), but it is, along with Hydra, among the longest that can be clearly traced at this time of year. **Eridanus** starts low in the sky, next to Rigel in Orion, and meanders upward in many curves, ending with the star Achernar, one of the brightest stars in the sky (see page 130).

In Greek mythology, Eridanus was the river into which Phaeton, the son of the sun god, plunged after his wild journey across the sky. Phaeton had asked his father, Helios, to let him drive the sun chariot, but Helios, worried about his son's safety, asked him to give up this wish. Phaeton, however, was determined to have his way and paid no heed to his father's warnings. The horses yanked the chariot around furiously as Phaeton was trying to steer them through the constellations of the zodiac. He became frightened and

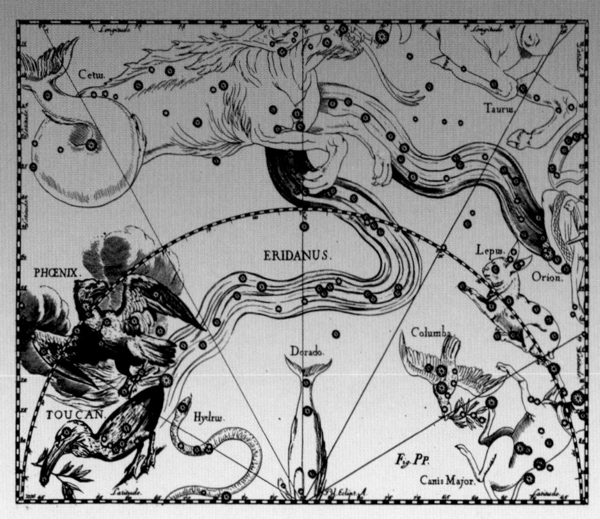

Eridanus

October 1, 11 P.M.
October 15, 10 P.M.
October 31, 9 P.M.

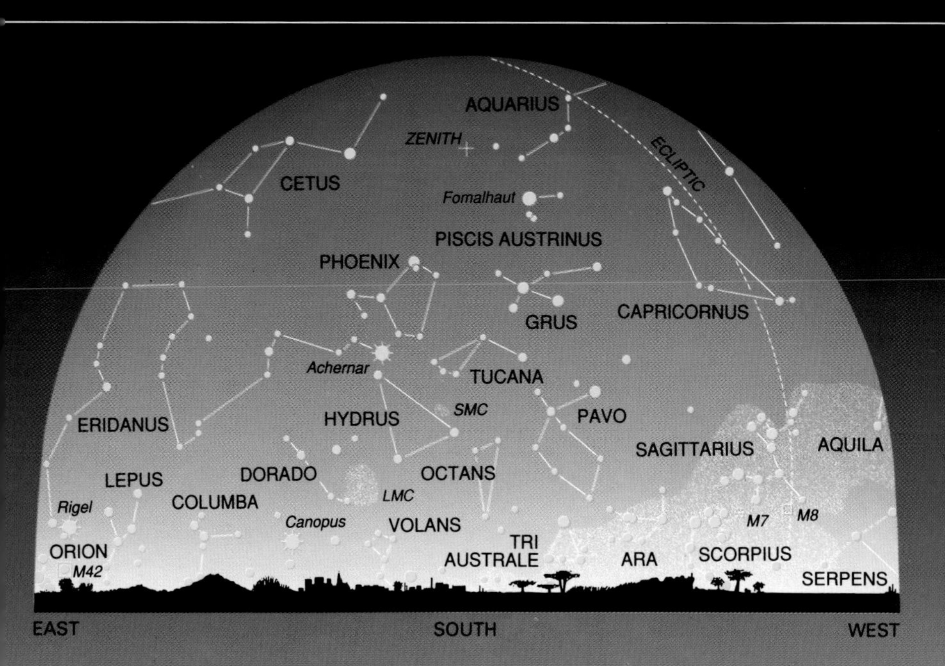

lost control of the horses. According to some accounts, the chariot's mad careening across the sky created the Milky Way (see also page 140). In any case, the fiery chariot came so close to Earth that it scorched the people of Ethiopia, leaving them black, and turned Libya into a red-hot desert. Eventually Phaeton was hurled into the river Eridanus, where the nymphs that lived there recovered and buried his body. The gods raised Eridanus to the heavens but not Phaeton.

Eridanus contains another star, apart from Achernar, which, though not especially conspicuous, frequently is mentioned when the question comes up whether there might be life on some planet in outer space. It is Epsilon Eridani, which bears a very close resemblance to our Sun. It is of about the same size and brightness and is a relatively close neighbor, a mere 11 light-years away. Because of these similarities, some people have suspected that Epsilon Eridani might have a planetary system similar to that of our Sun and that life might have evolved there as it did on Earth. That is why a project was initiated in 1960 of examining Epsilon Eridani and another look-alike of the Sun in Cetus intensively through a radio telescope, scanning them at a wavelength of 21 cm. The hope was that signals from an extraterrestrial civilization might be detected. The result was negative. To date, not a single signal has been received from Epsilon Eridani that gives any indication of artificial origin and might thus have been sent by creatures on a distant planet.

Next to Achernar is Phoenix, another constellation that was first introduced by Johann Bayer in 1603.

November

Star map S 11

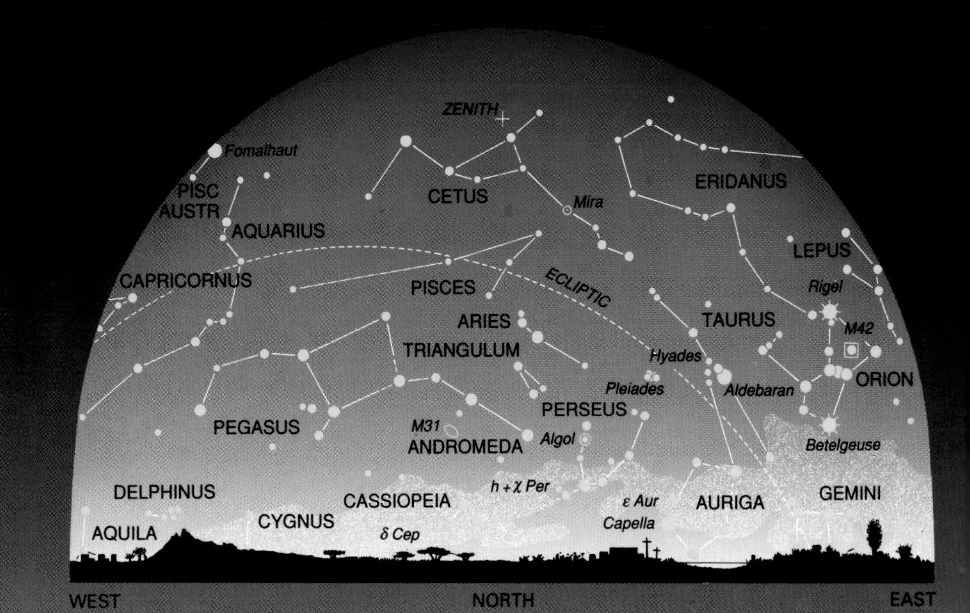

In the eastern sky, Orion is now particularly prominent. For stargazers in the Northern Hemisphere, this constellation has an unusual look because Orion seems to be standing on its head. Sirius, the brightest star of all, along with Canopus, is again in a good viewing position in the eastern sky, too. Sirius, Canopus, and Achernar stand in a more or less straight line and are of aid in getting oriented in the sky, especially at dusk, when not many other stars are yet visible.

Between Sirius and Orion a constellation with the unusual name Lepus (the Hare) shows up in the southeastern sky. This, too, is a classical constellation that already was known to the ancient civilizations. Its exact origin is unclear. Because its position is south of Orion, it might have represented some game animal that the heavenly hunter Orion was chasing.

Also in a good position for easy viewing are Tucana (the Toucan) and **Capricornus** (the Sea Goat). The latter is one of the constellations of the classical zodiac (see page 16), whereas the former is one of the constellations introduced by Johann Bayer. Although not particularly impressive otherwise, Tucana includes one of the notable objects in the southern sky, the Small Magellanic Cloud, which—along with the Large Magellanic Cloud in Dorado—is among the most striking celestial objects in the sky of the Southern Hemisphere.

The earliest modern mention of the Small Magellanic Cloud (see page 137) and of its larger namesake goes back to the first journey around the world in 1520, headed by Ferdinand Magellan. The clouds are named for the explorer, even though he was killed in the Philippines and his crew had to complete the journey without their leader. The Large and the Small Magellanic Clouds are the galaxies that lie closest to our solar system. They are irregular galaxies, that is, they have no particular shape. The Small Magellanic Cloud is somewhat farther away from us than the Large one. These

November 1, 11 P.M.
November 15, 10 P.M.
November 30, 9 P.M.

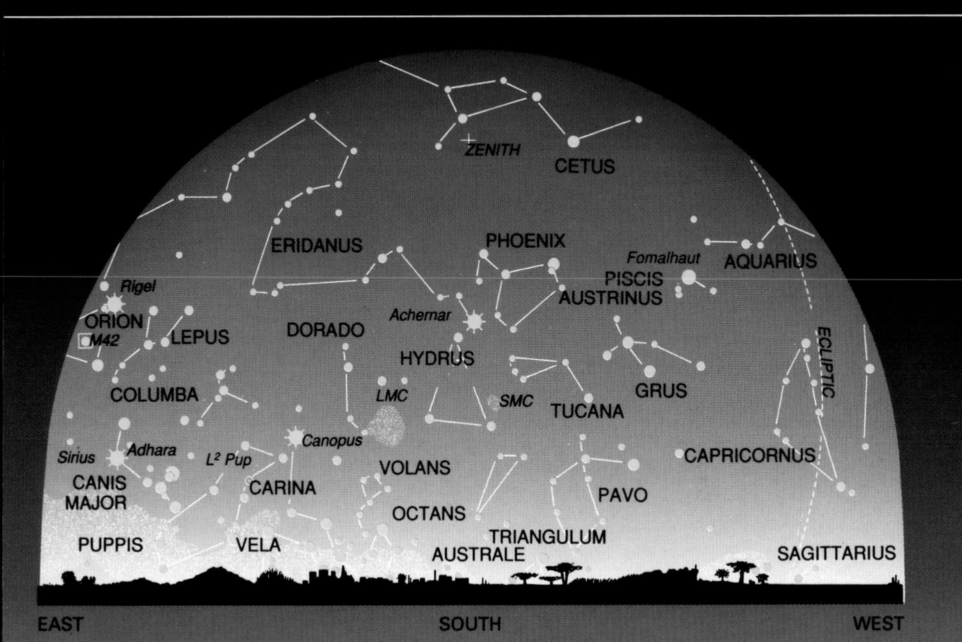

two galaxies are the prime objects of observation on which the telescopes located in the Southern Hemisphere are trained. Both galaxies are so close to the south celestial pole that they are never visible in the mid-latitudes of the north. Until about the 1950s the stars of the northern sky were studied much more thoroughly than those seen only in the Southern Hemisphere. This has changed dramatically within the last 30 years, as many observatories in the Northern Hemisphere, especially those in the United States and in Europe, have set up associate institutions on the southern continents. Thus, eight European nations have established the European Southern Observatory (ESO), which now maintains a huge observatory on La Sillia, a mountaintop in Chile.

November, finally, is the month of the Leonids, a large meteor shower. In the middle of the month, exceptionally large numbers of meteors appear to burst forth from within the constellation Leo (see page 60).

Capricornus

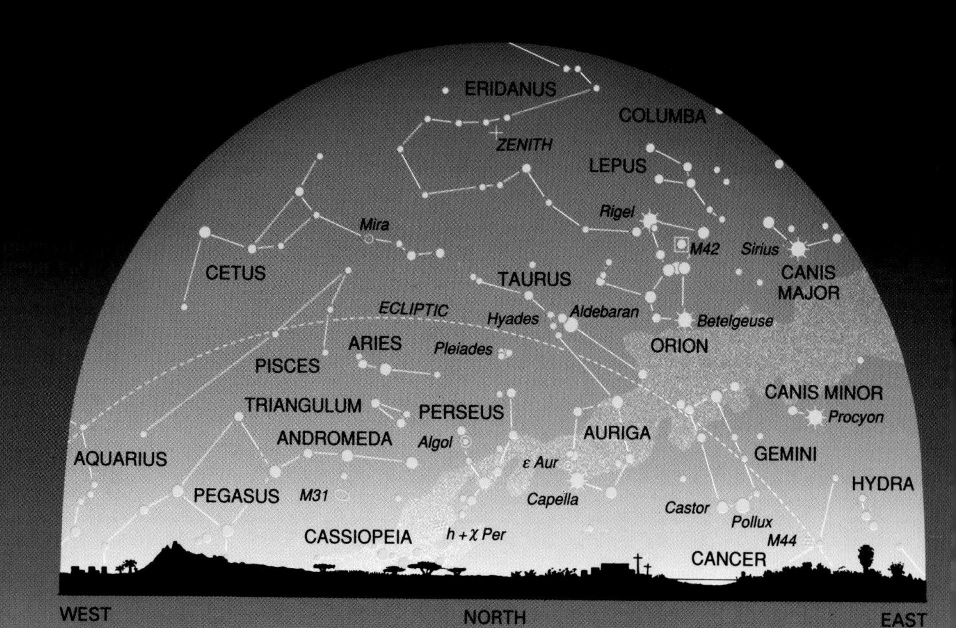

In studying the star maps for the month of December, one soon is struck by the fact that among the various types of names for constellations there is one group of animals that is unusually well represented, namely fish. In addition to Pisces, one of the constellations of the classical zodiac, we find Cetus (the Whale), Volans (the Flying Fish), Piscis Austrinus (the Southern Fish), and, finally, Dorado (the Swordfish). The last of these, Dorado, was invented and placed in the sky rather arbitrarily by Johann Bayer in 1603 and hardly would be worth special mention if it were not the site of the most glamorous object of the southern sky, the Large Magellanic Cloud. Even on nights when the full moon is out, the Large Magellanic Cloud can be seen plainly with the naked eye as a large, fuzzy patch of light, which, when viewed through binoculars or—to even more dramatic effect—through a telescope, reveals itself as a huge accumulation of innumerable stars.

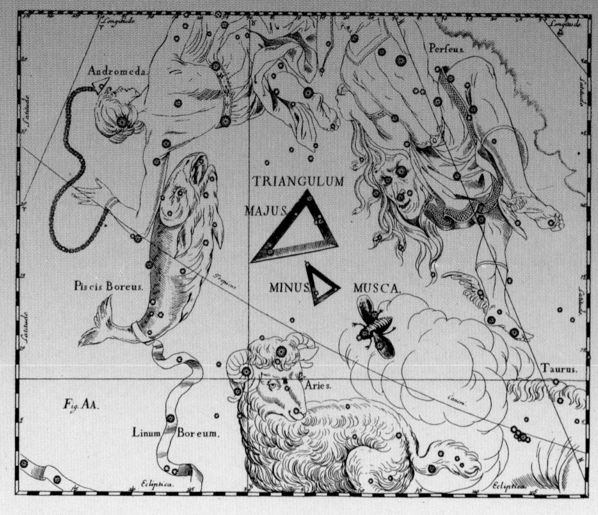

Triangulum

December 1, 11 P.M.
December 15, 10 P.M.
December 31, 9 P.M.

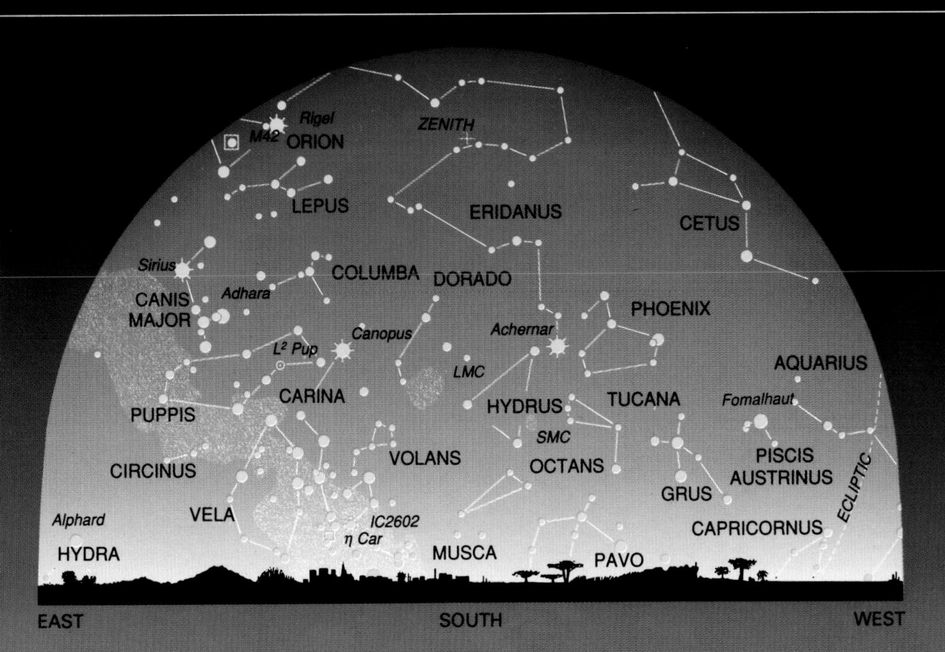

The Large Magellanic Cloud—named, like its smaller counterpart, for the explorer-navigator Magellan—is the galaxy that lies closest to Earth. It is only about 150,000 light-years distant, and astronomers estimate that it contains between 15 and 30 million stars.The Large Magellanic Cloud includes a number of nebulas, star clusters, and other objects, all of which stand out in beautiful colors in photographs of the cloud (see pages 152–53).

The Large Magellanic Cloud gained further fame in 1987, when a supernova was discovered there. This supernova, which technically is referred to as 1987 A, is the most recent supernova that could be seen by the unaided eye (see page 149). This supernova is particularly interesting because it was the first time that such an event could be recorded by the entire array of technological instruments at the disposal of modern astronomy. It therefore was observed by satellites as well as by all available telescopes on the southern half of the globe. An unusual signal originating in this distant astral explosion was recorded even in the deep shafts of mines, where the impact of neutrinos could be registered.

In the northern part of the sky there is a third distant galaxy that also is visible to the unaided eye. It is the Andromeda Galaxy, technically known as M 31. South of it lies the small constellation **Triangulum** (the Triangle), one of the few that goes back to ancient times and is named after an object rather than a mythological figure. In this case there is even a close resemblance between the object and its representation in the sky. Above Triangulum the ecliptic cuts across the sky in a low arc in December. The planets and the Moon, which travel along the ecliptic, are therefore also relatively low in the northern sky.

Celestial Events up to the Year 2010

The stars and the constellations formed by them always appear in the same place in the sky every year at the same time. They rise and set, thus demonstrating Earth's rotation around its own axis (see The Movement of Earth on page 8). The mindboggling, truly astronomical distances that separate them from Earth prevent us from perceiving their own motion—the so-called proper motion—at which they travel through space. Only heavenly bodies that are much closer to us than the stars display motion that can be discerned clearly from Earth as these bodies wander from one constellation to the next, continually changing position in relation to each other and to the stars that shine behind them. There are only a few of these movable heavenly bodies that are visible to the naked eye. But they constitute the most striking and spectacular celestial sights. Important events of this kind include the Sun in its relation to Earth; the Moon; eclipses resulting from the interplay of the Sun, the Moon, and Earth; and the major planets Mercury, Venus, Mars, Jupiter, and Saturn.

Unfortunately there is no way to predict the equally if not even more spectacular appearances of comets or shooting stars and of supernovas, the sudden, dramatic blazing up of stars (see pages 133 and 149). To keep abreast of these events, all you can do is follow reports in the daily press.

Predictable Facts Concerning the Sun, Earth, and the Beginning of the Seasons

Earth and the Sun are one inseparable unit in space. Without the Sun there could be no life on Earth, for only the Sun's energy, which has been radiating over Earth for more than five billion years, makes life possible here (concerning the Sun, see page 149).

The rising and setting of the Sun determines the rhythm of our days, and ever since man first appeared on Earth, the day has been the natural unit by which he has measured time. Earth's orbit around the Sun defines the length of the year, which is equivalent to 365¹/₄ revolutions of Earth around its axis. The motions of Earth and the Sun consequently form the basis of our calendar. Depending on the geographic latitude, the Sun rises and sets at different hours in the course of the year (see page 12), and depending on the season it is higher or lower in the sky and thus determines the length of the day (see pages 12–13).

The Beginning of the Seasons

Northern Hemisphere Southern Hemisphere	Spring Fall		Summer Winter		Fall Spring		Winter Summer	
2001	3/20	8:28 A.M.	6/21	2:34 A.M.	9/22	6:01 P.M.	12/21	2:19 P.M.
2002	3/20	2:16 P.M.	6/21	8:23 A.M.	9/22	11:55 P.M.	12/21	8:13 P.M.
2003	3/20	8:03 P.M.	6/21	2:11 P.M.	9/23	5:47 A.M.	12/22	2:02 A.M.
2004	3/20	1:49 A.M.	6/20	7:57 P.M.	9/22	11:29 A.M.	12/21	7:40 A.M.
2005	3/20	7:32 A.M.	6/21	1:48 A.M.	9/22	5:22 P.M.	12/21	1:35 P.M.
2006	3/20	1:24 P.M.	6/21	7:28 P.M.	9/22	11:04 P.M.	12/21	7:24 P.M.
2007	3/20	7:07 P.M.	6/21	1:09 P.M.	9/23	4:53 A.M.	12/22	1:09 A.M.
2008	3/20	12:48 A.M.	6/20	6:59 P.M.	9/22	10:46 A.M.	12/21	7:02 A.M.
2009	3/20	6:43 A.M.	6/21	12:43 A.M.	9/22	4:21 P.M.	12/21	12:46 P.M.
2010	3/20	12:32 P.M.	6/21	6:27 A.M.	9/22	10:11 P.M.	12/21	6:40 P.M.

Because of leap year, the dates marking the beginning of the seasons change in a four-year cycle.

The table on page 104 shows the exact time when spring, summer, fall, and winter begin in the years 2001 to 2010. In the Northern Hemisphere, the Sun is highest in the sky and shines the longest at the beginning of summer. After that it seems to sink lower and lower until it crosses to the south side of the celestial equator at the beginning of fall, at the so-called autumnal equinox. Three months later, at the onset of winter in the north and of summer in the south, the Sun reaches its lowest point in the sky of the Northern Hemisphere and the highest point in the sky of the southern half of the globe. This time is called a solstice because the sun "comes to a stop." It sinks no lower but starts rising higher again until, when it is spring in the Northern Hemisphere (fall on the southern half of the globe), it recrosses the equator, this time moving from south to north. This point of the year is called the vernal equinox, because once again—as at the autumnal equinox—day and night are almost exactly the same length.

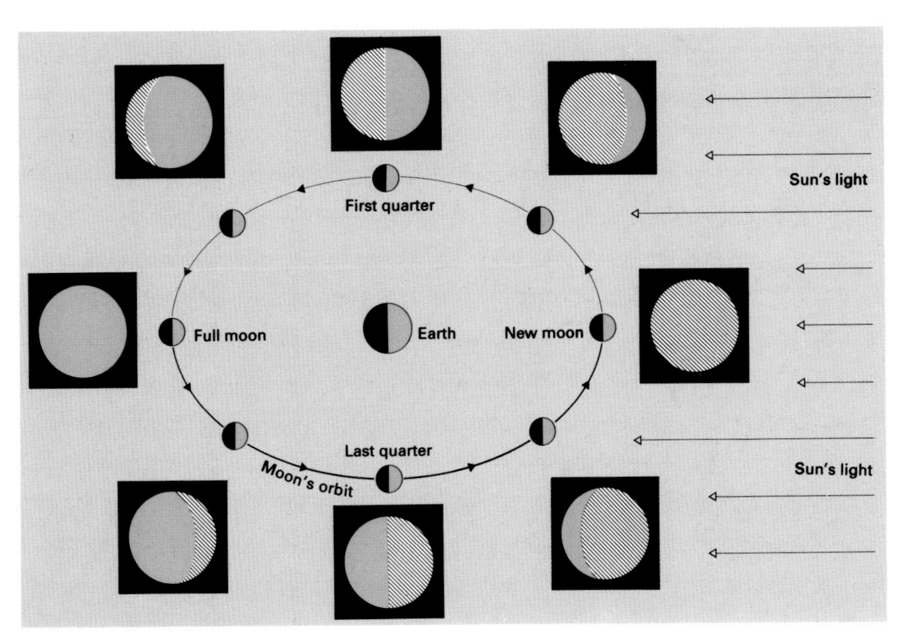

Depending on its position in relation to Earth and the Sun, the Moon, as we see it, is lit up differently by the Sun's light. The pictures in the black squares show how it appears to us in the sky.

All Phases of the Moon Can Be Predicted Accurately

The Moon is the heavenly body nearest to us—Earth's companion on the journey through space. It orbits continually around Earth and, along with Earth, travels around the Sun. The Moon is the only celestial body (apart from the Sun) that appears as a disk to the naked eye and on whose surface one clearly can see lighter and darker areas.

As it circles around Earth, the Moon is illuminated by the Sun—just like Earth—

and keeps changing in appearance, passing through the different lunar phases we all know. The diagram on page 105 shows how these phases come about. The whole cycle of the Moon's phases—the transformation from invisible new moon to a crescent in the evening sky that waxes gradually fatter night after night until half of the Moon's face is lit up at what is called the first quarter; the continued growth into a round disk when the Moon is full and stays in the sky all night; then the waning to the "last quarter," when the Moon shines only during the second half of the night; and, finally, the return to new moon—takes about 29½ days. This cycle and the time it takes to complete it is the basis of a further unit of measuring time, the lunar, or synodic, month. In many ancient cultures, the lunar month was a more important time unit than the year. In old myths it often determined the course of events, and in many ancient calendar systems the months corresponded exactly to the actual lunar phases. In our modern calendar, the months are merely divisions of the year and bear no relationship to the actual phases of the Moon.

But even today the changes of the lunar phases are one of the most prominent processes taking place in the sky, and all the special points of the lunar cycle—the dates of the two half moons, the full moon, and the new moon—are listed in the following table for the years 2001 to 2010.

Note: Lunar eclipses only occur during the full moon; solar eclipses occur only during the new moon.

Lunar Phases 2001–2010

In this chart, ● = New Moon, ◐ = First Quarter, ○ = Full Moon, ◑ = Last Quarter. Eclipse dates are given in bold numerals.

	JAN	FEB	MAR	APR	MAY	JUN	JUL	AUG	SEP	OCT	NOV	DEC
2001	2 ◐	1 ◐	2 ◐	1 ◐	7 ○	5 ○	**5** ○	4 ○	2 ○	2 ○	1 ○	7 ◑
	9 ○	8 ○	9 ○	7 ○	15 ◑	13 ◑	13 ◑	12 ◑	10 ◑	9 ◑	8 ◑	**14** ●
	16 ◑	14 ◑	16 ◑	15 ◑	22 ●	**21** ●	20 ●	18 ●	17 ●	16 ●	15 ●	22 ◐
	24 ●	23 ●	24 ●	23 ●	29 ◐	27 ◐	27 ◐	25 ◐	24 ◐	23 ◐	22 ◐	**30** ○
				30 ◐							30 ○	
2002	5 ◑	4 ◑	5 ◑	4 ◑	4 ◑	2 ◑	2 ◑	1 ◑	6 ●	6 ●	4 ●	**4** ●
	13 ●	12 ●	13 ●	12 ●	12 ●	**10** ●	10 ●	8 ●	13 ◐	13 ◐	11 ◐	11 ◐
	21 ◐	20 ◐	21 ◐	20 ◐	19 ◐	17 ◐	16 ◐	15 ◐	21 ○	21 ○	**19** ○	19 ○
	28 ○	27 ○	28 ○	26 ○	**26** ○	24 ○	24 ○	22 ○	29 ◑	29 ◑	27 ◐	26 ◑
								30 ◑				
2003	2 ●	1 ●	2 ●	1 ●	1 ●	7 ◐	6 ◐	5 ◐	3 ◐	2 ◐	**8** ○	8 ○
	10 ◐	9 ◐	11 ◐	9 ◐	9 ◐	14 ○	13 ○	11 ○	10 ○	10 ○	16 ◑	16 ◑
	18 ○	16 ○	18 ○	16 ○	**16** ○	21 ◑	21 ◑	19 ◑	18 ◑	18 ◑	**23** ●	23 ●
	25 ◑	23 ◑	24 ◑	23 ◑	22 ◑	29 ●	29 ●	27 ●	25 ●	25 ●	30 ◐	30 ◐
					30 ●					31 ◐		

Lunar Phases 2001–2010

	JAN	FEB	MAR	APR	MAY	JUN	JUL	AUG	SEP	OCT	NOV	DEC
2004	7 ○	6 ○	6 ○	5 ○	**4** ○	2 ○	2 ○	7 ◑	6 ◑	6 ◑	5 ◑	4 ◑
	15 ◑	13 ◑	13 ◑	11 ◑	11 ◑	9 ◑	9 ◑	15 ●	14 ●	**14** ●	12 ●	11 ●
	21 ●	20 ●	20 ●	**19** ●	18 ●	17 ●	17 ●	23 ◐	21 ◐	20 ◐	19 ◐	18 ◐
	29 ◐	27 ◐	28 ◐	27 ◐	27 ◐	25 ◐	24 ◐	29 ○	28 ○	**27** ○	26 ○	26 ○
							31 ○					
2005	3 ◑	2 ◑	3 ◑	1 ◑	1 ◑	6 ●	6 ●	4 ●	3 ●	**3** ●	1 ●	1 ●
	10 ●	8 ●	10 ●	**8** ●	8 ●	14 ◐	14 ◐	12 ◐	10 ◐	10 ◐	8 ◐	8 ◐
	17 ◐	15 ◐	17 ◐	16 ◐	16 ◐	21 ○	21 ○	19 ○	17 ○	**17** ○	15 ○	15 ○
	25 ○	23 ○	25 ○	**24** ○	23 ○	28 ◑	27 ◑	26 ◑	25 ◑	24 ◑	23 ◑	23 ◑
					30 ◑							30 ●
2006	6 ◐	5 ◐	6 ◐	5 ◐	5 ◐	3 ◐	3 ◐	2 ◐	**7** ○	6 ○	5 ○	4 ○
	14 ○	12 ○	**14** ○	13 ○	13 ○	11 ○	10 ○	9 ○	13 ◑	13 ◑	12 ◑	12 ◑
	22 ◑	21 ◑	22 ◑	20 ◑	20 ◑	18 ◑	17 ◑	15 ◑	**22** ●	22 ●	20 ●	20 ●
	29 ●	27 ●	**29** ●	27 ●	27 ●	25 ●	24 ●	23 ●	30 ◐	29 ◐	28 ◐	27 ◐
								31 ◐				
2007	3 ○	2 ○	**3** ○	2 ○	2 ○	8 ◑	7 ◑	5 ◑	1 ◑	3 ◑	1 ◑	1 ◑
	11 ◑	10 ◑	11 ◑	10 ◑	9 ◑	14 ●	13 ●	12 ●	**11** ●	11 ●	9 ●	9 ●
	18 ●	17 ●	**18** ●	17 ●	15 ●	22 ◐	22 ◐	20 ◐	19 ◐	19 ◐	17 ◐	17 ◐
	25 ◐	24 ◐	25 ◐	24 ◐	23 ◐	30 ○	29 ○	**28** ○	26 ○	25 ○	24 ○	23 ○
					31 ○							31 ◑
2008	8 ●	**6** ●	7 ●	5 ●	5 ●	3 ●	1 ●	**1** ●	7 ◐	7 ◐	5 ◐	5 ◐
	15 ◐	13 ◐	14 ◐	12 ◐	11 ◐	10 ◐	9 ◐	8 ◐	15 ○	14 ○	13 ○	12 ○
	22 ○	**20** ○	21 ○	20 ○	19 ○	18 ○	18 ○	**16** ○	22 ◑	21 ◑	19 ◑	19 ◑
	30 ◑	28 ◑	29 ◑	28 ◑	27 ◑	26 ◑	25 ◑	23 ◑	29 ●	28 ●	28 ●	27 ●
								30 ●				
2009	4 ◐	2 ◐	4 ◐	2 ◐	1 ◐	7 ○	**7** ○	5 ○	4 ○	4 ○	2 ○	2 ○
	10 ○	**9** ○	10 ○	9 ○	8 ○	15 ◑	15 ◑	13 ◑	11 ◑	11 ◑	9 ◑	8 ◑
	17 ◑	15 ◑	18 ◑	17 ◑	17 ◑	22 ●	**21** ●	20 ●	18 ●	18 ●	16 ●	16 ●
	26 ●	24 ●	26 ●	24 ●	24 ●	29 ◐	28 ◐	27 ◐	25 ◐	25 ◐	24 ◐	24 ◐
					30 ◐							**31** ○
2010	7 ◑	5 ◑	7 ◑	5 ◑	5 ◑	4 ◑	4 ◑	2 ◑	1 ◑	7 ●	5 ●	5 ●
	15 ●	13 ●	15 ●	15 ●	13 ●	12 ●	**11** ●	9 ●	8 ●	14 ◐	13 ◐	13 ◐
	23 ◐	21 ◐	23 ◐	21 ◐	20 ◐	18 ◐	18 ◐	16 ◐	15 ◐	22 ○	21 ○	**21** ○
	30 ○	28 ○	29 ○	28 ○	27 ○	**26** ○	25 ○	24 ○	23 ○	30 ◑	28 ◑	27 ◑
									30 ◑			

Dates, Duration, and Visibility of Solar Eclipses

The Moon's circling around Earth is responsible for what are probably the most dramatic celestial events: the solar and lunar eclipses. A solar eclipse occurs when the Moon stands between Earth and the Sun, blocking the sunlight and casting its shadow across Earth.

A lunar eclipse occurs when Earth stands between the Sun and the Moon and casts its shadow across the surface of the Moon (see diagram below).

From these two statements an important law can be deduced, namely, that solar eclipses can occur only during new moon (the only time when the Moon stands between Earth and the Sun) and lunar eclipses can occur only during full moon (the only time when the Moon is at the exact opposite point of the sky from the Sun, with Earth on the line between them). You can see very clearly that this law holds if you look at the table of the lunar phases on pages 106 and 107, where the solar and lunar eclipses for the years 2001 to 2010 are entered. (The exact dates are printed in boldface.)

Both the shadow of the Moon and the shadow of Earth are made up of two parts. Because both bodies are spherical, they cast conical shadows that consist of an outer and an inner cone. The inner cone points away from the Sun (or Earth), whereas the outer cone points toward the Sun (or Earth). The resulting shadows are called, respectively, the umbra and the penumbra. The interplay between them results in different kinds of eclipses. The umbra of the Moon is almost exactly the same length as the mean distance between Earth and the Moon, that is,

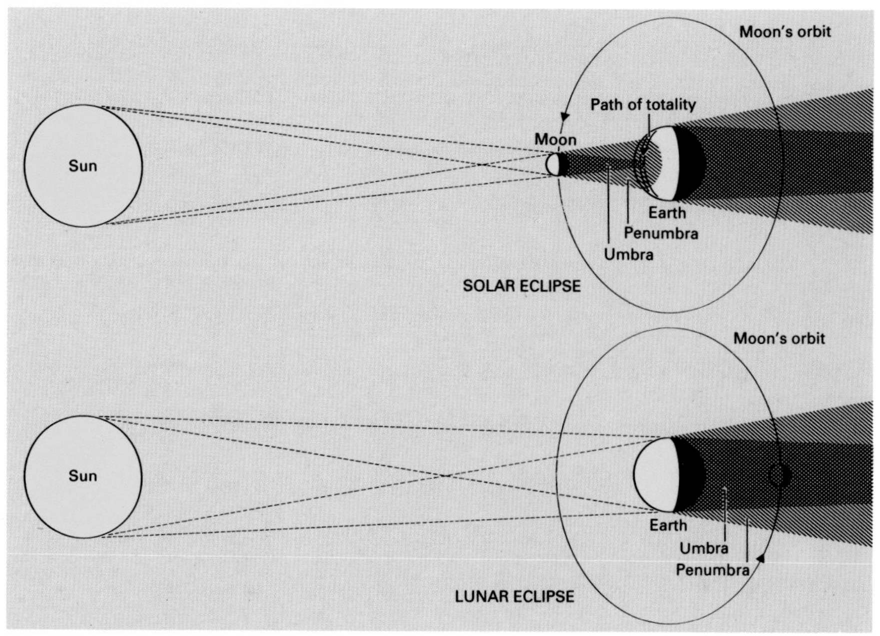

During a solar eclipse, the Moon's shadow covers Earth. During a lunar eclipse, Earth's shadow covers the Moon.

approximately 240,000 miles (380,000 km). Consequently the tip of the umbra just barely reaches the surface of Earth, where it covers an area of maximally 170 miles (273 km) in diameter. Around this area there is a considerably larger circle (up to 4,400 miles or 7,000 km across!) that is reached by the penumbra. In the umbral area the Sun appears completely covered (see photo, page 110), whereas in the area of the penumbra it is only partially covered (see photos on this page).

While the Moon's shadow falls on Earth's surface, both Earth and the Moon continue to move, Earth rotating around its axis from west to east, and the Moon traveling along in the same direction. The result of this motion is that the umbra as well as the penumbra flit across Earth's surface at great speed (about 1,500 miles or 2,400 km per hour!), always from west to east. (The exact paths they take across the globe's surface are indicated on the maps of the most important solar eclipses reproduced on the flap and the inside front cover.)

The dates of all eclipses through the year 2010 are listed in the table on pages 110 and 111. Included there are two other kinds of solar eclipses: partial eclipses and annular eclipses. If the Moon is too far north or south of the Sun at new moon, only the penumbra may reach Earth's surface, resulting in a partial solar eclipse. Such eclipses are visible only in the extreme northern and southern latitudes, that is in the Arctic and nearby regions and on Antarctica. During an annular eclipse, the Moon's umbra hits the surface of Earth, just as during a total eclipse. But because the distances between Earth and the Sun and between Earth and the Moon fluctuate, it sometimes happens that the Moon's umbra doesn't quite reach Earth's surface. If one then stands on a point that is directly in line with the umbra's tip, one witnesses the strange sight of the Moon's disk blocking the Sun but not covering it

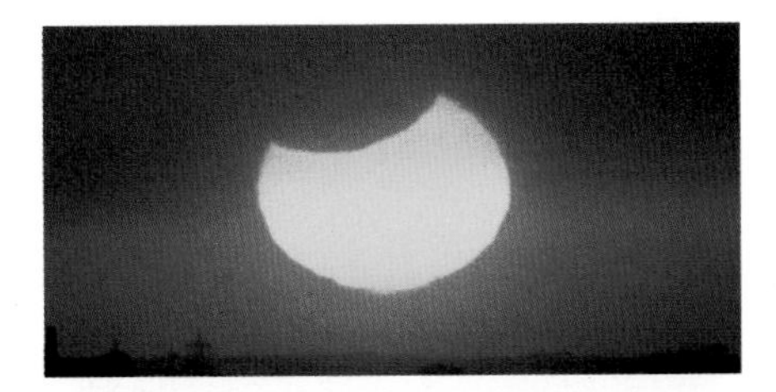

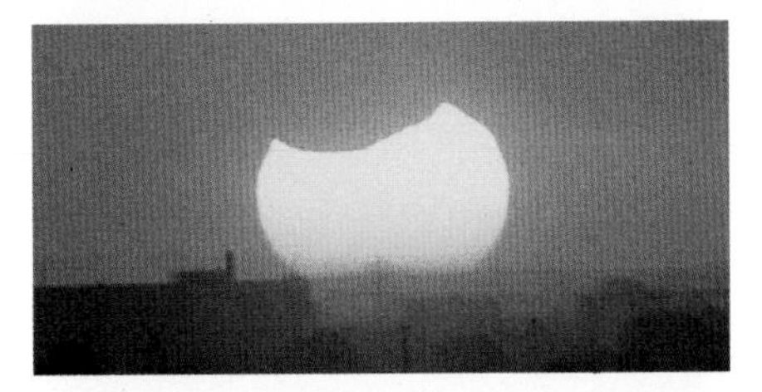

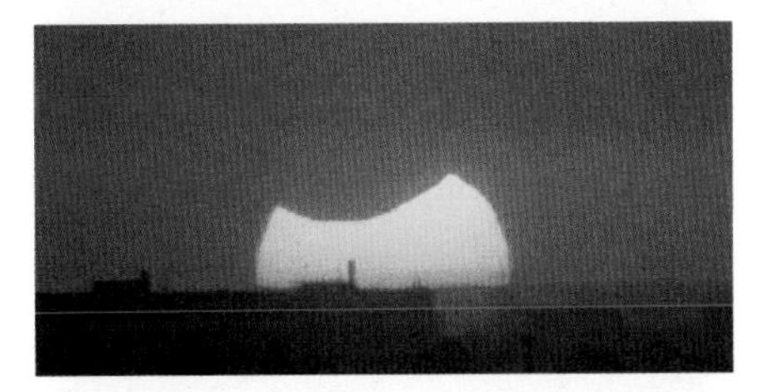

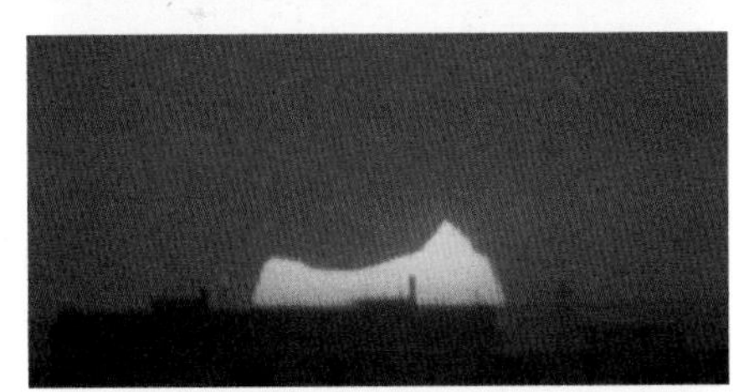

A rare celestial event: The partially eclipsed Sun drops below the horizon on July 20, 1982, as seen in Hamburg, Germany.

completely, as it would during a total eclipse. Instead, a narrow ring of the Sun's surface still shows around the dark moon. That is why this kind of eclipse is called annular (from Latin *annulus*, ring).

Total and annular solar eclipses (less so the partial ones) surely constitute the most dramatic celestial phenomena of all. It is only during a total solar eclipse that the solar corona in its full extent can be seen. The corona (a garland or crown of light surrounding the Sun) is the Sun's atmosphere, which consists of hot, highly dissipated gases (see page 149) and extends several million miles into space. It appears in ever changing forms from one eclipse to another and offers a spectacular scene. The sudden darkness that falls over Earth in the middle of the day, the emergence of the brighter stars in the sky, and the reactions of natural creatures and organisms for whom this sudden darkness comes as a complete surprise—all this turns a total solar eclipse into an unforgettable experience.

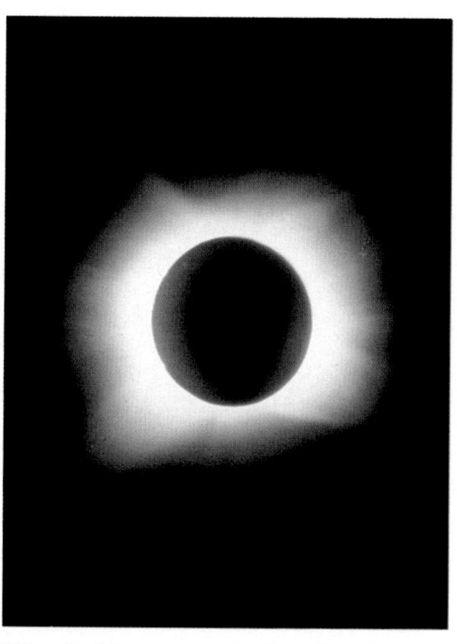

When the Moon covers the Sun, the solar corona becomes visible (Java, June 11, 1983).

Because the Moon's umbra narrows down to such a small area by the time it reaches Earth's surface, total solar eclipses are extremely rare in any one particular country. Partial solar eclipses are considerably more common because the Moon's penumbra measures about 4,400 miles (7,000 km) in diameter, compared to a maximum of 170 miles (273 km) for the umbra.

Solar Eclipses: When and Where They Are Visible

Date	Type	Peak of Eclipse	Duration of Peak Obscurity	Maximum Extent of Solar Disk Obscured
Jun. 21, 2001	t	7:03 A.M.	4 min, 56 sec	105%
Visibility: See front of book				
Dec. 14, 2001	a	3:51 P.M.	3 min, 54 sec	97%
Visibility: See front of book				
Jun. 10, 2002	a	6:43 P.M.	1 min, 13 sec	99%
Visibility: See front of book				
Dec. 4, 2002	t	2:30 A.M.	2 min, 4 sec	102%
Visibility: South Africa, Indian Ocean, Australia				
May 30, 2003	a	11:07 P.M.	3 min, 37 sec	94%
Visibility: Iceland, Arctic				
Nov. 23, 2003	t	5:48 P.M.	1 min, 57 sec	104%
Visibility: Antarctica				

Solar Eclipses 2001–2010

Solar Eclipses: When and Where They Are Visible

Date	Type	Peak of Eclipse	Duration of Peak Obscurity	Maximum Extent of Solar Disk Obscured
Apr. 19, 2004	p	8:33 A.M.	—	74%
Visibility: South Pacific, South Africa				
Oct. 13, 2004	p	9:58 P.M.	—	93%
Visibility: Asia, North Pacific				
Apr. 8, 2005	a–t	3:35 P.M.	42 sec	100%
Visibility: See front of book				
Oct. 3, 2005	a	5:30 A.M.	4 min, 32 sec	96%
Visibility: See front of book				
Mar. 29, 2006	t	5:10 A.M.	4 min, 7 sec	105%
Visibility: See front of book				
Sep. 22, 2006	a	6:39 A.M.	7 min, 9 sec	94%
Visibility: Northeast of South America, Atlantic				
Mar. 18, 2007	p	9:31 P.M.	—	87%
Visibility: Asia, western Alaska				
Sep. 11, 2007	p	7:30 A.M.	—	75%
Visibility: Southern South America, southern Atlantic Ocean, Antarctica				
Feb. 6, 2008	a	10:54 P.M.	2 min, 14 sec	97%
Visibility: South Pacific Ocean, Antarctica				
Aug. 1, 2008	t	5:20 A.M.	2 min, 28 sec	104%
Visibility: See front of book				
Jan. 26, 2009	a	2:57 A.M.	7 min, 56 sec	93%
Visibility: South Atlantic, Indian Ocean				
Jul. 21, 2009	t	9:34 P.M.	6 min, 40 sec	108%
Visibility: Asia, Pacific Ocean				
Jan. 15, 2010	a	2:05 A.M.	11 min, 10 sec	92%
Visibility: Africa, Indian Ocean				
Jul. 11, 2010	t	1:32 P.M.	5 min, 20 sec	106%
Visibility: Pacific Ocean, South America				

p = partial; t = total; a = annular; min = minutes, sec = seconds

Dates, Duration, and Visibility of Lunar Eclipses

Unlike solar eclipses, lunar eclipses can be observed with considerably greater frequency in any given country. The simple reason for this is that every lunar eclipse is visible on the entire dark side of Earth. However, fascinating lunar eclipses are rarer than solar eclipses, as the table below, which lists all the observable lunar eclipses over a ten-year period, shows. It lists only 15 lunar eclipses, whereas the table on pages 110 and 111 lists 20 solar eclipses for the years 2001 to 2010.

Earth, like the Moon, projects a shadow into space that consists of an umbra and a penumbra, but only the umbra is of interest in connection with lunar eclipses. When the Moon traverses Earth's penumbra, this is difficult to see from Earth. That is why the table below includes only eclipses during which the Moon travels through Earth's umbra.

When the fully lit disk of the Moon (the full moon) moves completely into Earth's umbra, a total eclipse occurs; when it moves past the umbra's center to the north or south and the umbra covers it only partially, it is a

Lunar Eclipses: When and Where They Are Visible

Date	Type	Beginning	End	Extent
Jan. 9, 2001	t	1:41 P.M.	4:56 P.M.	118%
Visibility: See back of book				
Jul. 5, 2001	p	8:37 A.M.	11:15 A.M.	49%
Visibility: South Africa, Asia, Australia, Pacific Ocean, extreme western United States, Mexico				
May 15–16, 2003	t	9:04 P.M.	12:17 A.M.	113%
Visibility: See back of book				
Nov. 8, 2003	t	6:33 P.M.	10:03 P.M.	102%
Visibility: See back of book				
May 4, 2004	t	1:47 P.M.	5:09 P.M.	130%
Visibility: South America, Atlantic Ocean, Africa, western Asia, Indian Ocean				
Oct. 27, 2004	t	8:14 P.M.	11:51 P.M.	131%
Visibility: See back of book				
Oct. 17, 2005	p	6:32 A.M.	7:27 A.M.	6%
Visibility: North America, Pacific Ocean, Australia, Asia				
Sep. 7, 2006	p	1:07 P.M.	2:36 P.M.	17%
Visibility: Africa, Europe, Asia, Indian Ocean, Atlantic Ocean				
Mar. 3, 2007	t	4:32 P.M.	8:11 P.M.	123%
Visibility: See back of book				
Aug. 28, 2007	t	3:49 A.M.	7:20 A.M.	147%
Visibility: See back of book				
Feb. 20–21, 2008	t	8:45 P.M.	12:08 A.M.	111%
Visibility: See back of book				
Aug. 16, 2008	p	2:34 P.M.	5:41 P.M.	81%
Visibility: South America, Pacific Ocean, Africa, Europe, Asia				
Dec. 31, 2009	p	1:55 P.M.	2:51 P.M.	7%
Visibility: Atlantic Ocean, Africa, Europe, Asia, Indian Ocean				
Jun. 26, 2010	p	5:17 A.M.	7:58 A.M.	53%
Visibility: North and South America, eastern Asia, Pacific Ocean, Australia				
Dec. 21, 2010	t	1:32 A.M.	4:58 A.M.	125%
Visibility: See back of book				

p = partial; t = total

partial eclipse. (The exact extent of almost all the total lunar eclipses up to and including the year 2010 are shown on the back flap and inside cover.) Because Earth rotates, there is always a wide border zone in which the Moon either rises or sets during an eclipse.

To a hypothetical observer on the Moon, a lunar eclipse would appear the way a solar eclipse appears to us. For just as Earth plunges into the Moon's shadow during a solar eclipse, the Moon is engulfed by Earth's shadow as Earth moves between it and the Sun. Seen from Earth, the brightly lit disk of the Moon begins to darken more and more until it gradually is extinguished. But—and this is the crucial and really intriguing part of it all—the Moon's disk does not fade away completely but remains faintly visible through Earth's shadow even at the height of a total eclipse. This is because of Earth's atmosphere, which refracts the Sun's light and redirects some of it into Earth's shadow, where it strikes the Moon's surface and is reflected by it. Because this is true especially of the red spectrum of the Sun's light, the Moon often glows in a dark, coppery red during a total eclipse. However, this varies from eclipse to eclipse; the color depends primarily on the dust particles in Earth's atmosphere (see photo on front cover of book).

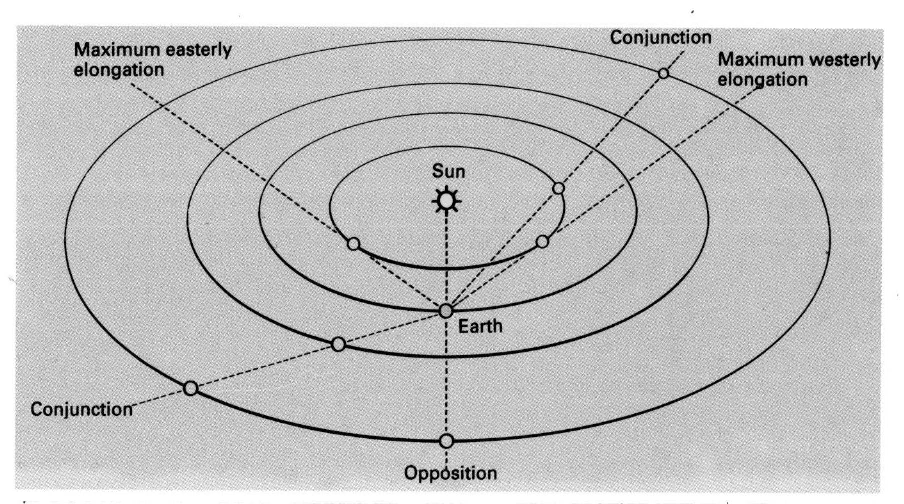

The planets can be aligned differently with Earth and the Sun. The different positions are called opposition, conjunction, and elongation (see page 114).

The Planets Appear in the Constellations of the Zodiac

The planets are lit up by the Sun and reflect the light toward Earth. Because of their proximity to Earth, they appear to move. This movement is always along the ecliptic, that is, the path the Sun seems to take across the sky in the course of the year. The ecliptic is entered in the star maps as a dotted line, so that it will be easy to find the planets in the sky. Planets are always very bright and they are always found along the ecliptic (see page 15). Their various positions in relation to Earth and the Sun are shown in the diagram above.

The famous constellations of the zodiac are lined up along the ecliptic (see pages 15–16). The planets therefore can show up only within these 13 constellations, which makes it even easier to spot them.

The appearance of the planets depends on whether they are closer to or farther away from the Sun than Earth (see diagram on page 17).

Mercury and Venus, the so-called inferior planets, are closer to the Sun than Earth is; the "superior" planets Mars, Jupiter, and Saturn are farther away.

Because Mercury and Venus orbit so closely to the Sun, they never can move very far away from it. They therefore are visible only in the evening in the western sky shortly after sunset or early in the morning in the eastern sky just before sunrise. The best position to observe them in is when their distance from the Sun, which is measured in degrees of an angle, is greatest. (The times when this is the case—the so-called *maximum elongation*—are given for Mercury in the table on pages 114 and 115 and for Venus in the table on page 116.)

Mars, Jupiter, and Saturn move around the Sun outside of Earth's orbit. In Earth's sky they can be opposite to the Sun, in a position where they are said to be at *opposition*. The times when this happens are given in the table on page 117. When one of these three planets is at opposition, it means that it is in its best position for observation because the planet is then in the sky all night. It rises approximately at sunset and sets when the Sun rises again.

The planets move across the sky at different rates. Mercury is the fastest; Saturn, the slowest. That is why their positions relative to each other keep changing. Often two or more planets converge in one of the constellations of the zodiac and then are aligned with each other exactly in a north-south direction. This is called a *conjunction* (see diagram on page 113). Planetary conjunctions are important celestial events; the sky appears different and striking because several planets are added to these constellations. (The most interesting planetary conjunctions for the years 2001 to 2010 are shown in small diagrams accompanying the descriptions of all the celestial events for that time span, starting on page 119.)

Mercury Appears Only Briefly

Mercury is the most difficult planet to observe both in the evening and in the morning. (The table lists only the times when this planet is particularly far east of the Sun and is thus visible in the evening.) The tropical countries are the best place from which to watch Mercury because dusk is so brief there and the planet is still relatively high above the horizon in the west when darkness falls. In zones farther north or south—that is, especially for the star map series N I (50 degrees north)—Mercury can be seen only with difficulty. The table for Mercury therefore indicates whether the planet is in an unfavorable position for observation anywhere on Earth or whether it can be seen better from the Northern or the Southern Hemisphere. These remarks are of no consequence for the equatorial regions.

Mercury 2001–2010

Year	Evening Sky (Constellation)	Visibility
2001	February (Aquarius)	unfavorable
	May–June (Taurus)	south
	September–October (Leo/Virgo)	south
2002	January (Capricornus)	north
	May (Taurus)	north
	August–September (Leo/Virgo)	south
2003	January (Sagittarius)	unfavorable
	April (Pisces)	north
	July–August (Leo/Virgo)	south
	November–December (Scorpius/Sagittarius)	south

Mercury 2001–2010

Year	Evening Sky (Constellation)	Visibility
2004	March–April (Aries/Taurus)	north
	July–August (Cancer/Leo)	south
	October–December (Libra/Scorpius)	south
2005	March (Pisces)	north
	June–July (Gemini/Cancer)	south
	October–November (Libra/Scorpius)	south
2006	February (Aquarius)	north
	June (Taurus/Gemini)	south
	October (Virgo/Libra)	south
2007	February (Aquarius)	north
	May–June (Taurus/Gemini)	north
	August–October (Leo/Virgo)	south
2008	January (Capricornus)	north
	May (Taurus)	north
	August/September (Leo/Virgo)	south
	December (Sagittarius)	south
2009	January (Capricornus)	south
	April–May (Aries/Taurus)	north
	August–September (Leo/Virgo)	south
	November–December (Scorpius/Sagittarius)	south
2010	April (Pisces)	north
	July–August (Cancer/Leo)	south
	November–December (Libra/Scorpius)	south

There Mercury is clearly visible in the evening in the west at any of the times indicated; however, one always has to look for it low on the horizon. (See also page 137.)

Mercury's Transits

When Mercury passes between Earth and the Sun, it happens once in a while that the planet passes directly in front of the solar disk. This is a kind of mini-eclipse with Mercury instead of the Moon covering part of the Sun. However, Mercury is so far away from Earth that it appears only as a tiny black dot against the Sun's disk, which is 150 times larger than that of the planet. During this decade, two of these rare Mercury transits will occur, but watching them is both difficult and dangerous. The only truly safe way is to project the Sun's image through a small telescope or through small binoculars onto a sheet of white paper and follow Mercury's transit there.

On May 7, 2003, in the early morning hours, around 3:00 A.M. Eastern Time, a Mercury transit of the Sun will occur. The second one this decade will take place on November 8, 2006, at around 4:45 P.M.

Venus, the Morning and Evening Star

Unlike Mercury, Venus can move much farther away from the Sun and consequently can stay in the sky much later in the evening after sunset and appear again much earlier before sunrise. It is also the brightest heavenly object apart from the Sun and the Moon and is therefore often visible before sunset, when it is still daylight. The table (on page 116) includes not only evening appearances but also those in the morning because Venus is often the only celestial body to be seen on one's way to work or when one first looks at the sky in the morning. Its still, white luminosity makes it a point of light that can't be missed. (See also page 151.)

Venus 2001–2010

Year	Evening Star (maximum elongation from Sun and constellation during the month of maximum elongation)	Morning Star
2001	January–March (1/19, Aquarius, Pisces)	April–December (6/12, Taurus, Gemini, Cancer)
2002	January–October	November–December (8/20, Cancer, Leo, Virgo)
2003	August–December	January–July (1/11, Libra, Scorpius)
2004	January–May (3/31, Aries, Taurus)	June–December (8/21, Gemini, Cancer)
2005	May–December (11/5, Scorpius, Sagittarius)	January–February
2006	January December	January–September (3/26, Capricornus, Aquarius)
2007	January–August (6/8, Gemini, Cancer)	September–December (10/29, Leo, Virgo)
2008	July–December	January–April
2009	January–March (1/15, Aquarius, Pisces)	April–November (6/7, Pisces, Aries)
2010	February–October (8/20, Leo, Virgo)	November–December

Transits of Venus are much more rare than those of Mercury, but there will be one this decade also. It will be the first since 1882, occurring on June 8, 2004. For about 20 minutes just after midnight Eastern Time, a tiny speck (the disk of Venus) will be seen traversing the Sun from Europe, Africa, and Asia.

Mars, the Red Planet

The planet Mars can come very close to Earth. It is second nearest to Earth, after Venus. Its visibility therefore follows an irregular pattern, too. There are years when it is never found at opposition, and there are years when it does not appear in the evening sky at all but only in the morning (see table on page 117). Mars glows with a reddish light; it is always fainter than Venus and almost always fainter than Jupiter as well, but generally exceeds even the most luminous "fixed" stars in brightness. Its light has a reddish cast to it that is caused by extensive deserts that cover huge areas of its surface. (See also page 137.)

Jupiter in a Yellowish Light

Jupiter, which is by far the largest planet of the solar system, shines with a clearly visible, yellowish light. It is very bright, constituting the second most luminous celestial object, after Venus (and excepting the Sun and the Moon). Only on very rare occasions does Mars outshine it for a few weeks during an opposition with exceptionally favorable conditions (this will not, however, happen at all during this decade). Jupiter's orbit around the Sun takes about 12 years, and it therefore moves along the sky from one zodiacal constellation to the next with each opposition, as the table on page 117 clearly shows. (See also page 136.)

Saturn, the Planet in White

Saturn is the slowest moving of the planets that can be observed with the unaided eye. It is also the least bright, although it is still on a par with the brightest stars in the sky. In looking for Saturn, one is helped not only by the fact that the planet always appears exactly on the ecliptic but also—and even more—by its steady light that barely flickers at all (see page 17 and the photo on pages 6–7). The famous rings of Saturn can be seen only with the aid of a telescope with a magnification of at least 40. In the year 2009, the rings will not be visible for a while because they are tipped toward Earth in such a way that they are on edge to our line of sight, thus, making them too narrow to be seen. Unfortunately, this will happen during a time when Saturn appears too close to the Sun to be seen. (See also page 146.)

Mars 2001–2010

Year	Opposition	Constellation at Opposition	Visible in the Evening Sky
2001	June 14	Sagittarius	May–December
2002	—	—	January–May (Aquarius–Taurus)
2003	August 29	Aquarius	July–December
2004	—	—	January–August (Pisces–Leo)
2005	November 7	Aries	September–December
2006	—	—	January–August (Aries–Leo)
2007	December 25	Gemini	October–December
2008	—	—	January–September (Gemini–Virgo)
2009	—	—	(only in the morning before sunrise)
2010	January 30	Cancer	January–December

Jupiter 2001–2010

Year	Opposition	Constellation at Opposition	Visible in the Evening Sky
2001	—	—	January–May (Taurus) November–December (Gemini)
2002	January 1	Gemini	January–July/November–December
2003	February 2	Cancer	January–August
2004	March 4	Leo	January–August
2005	April 4	Virgo	February–October
2006	May 5	Libra	March–November
2007	June 6	Ophiuchus	May–December
2008	July 9	Sagittarius	June–December
2009	August 15	Capricornus	July–December
2010	September 22	Pisces	January–February/August–December

Saturn 2001–2010

Year	Opposition	Constellation at Opposition	Visible in the Evening Sky
2001	December 4	Taurus	September–December
2002	December 18	Taurus	January–May/October–December
2003	—	—	January–June/November–December (Gemini)
2004	January 1	Gemini	January–June/October–December
2005	January 14	Gemini	January–July/November–December
2006	January 28	Cancer	January–July/December
2007	February 11	Leo	January–August/December
2008	February 25	Leo	January–August
2009	March 9	Leo	January–September
2010	March 22	Virgo	January–September

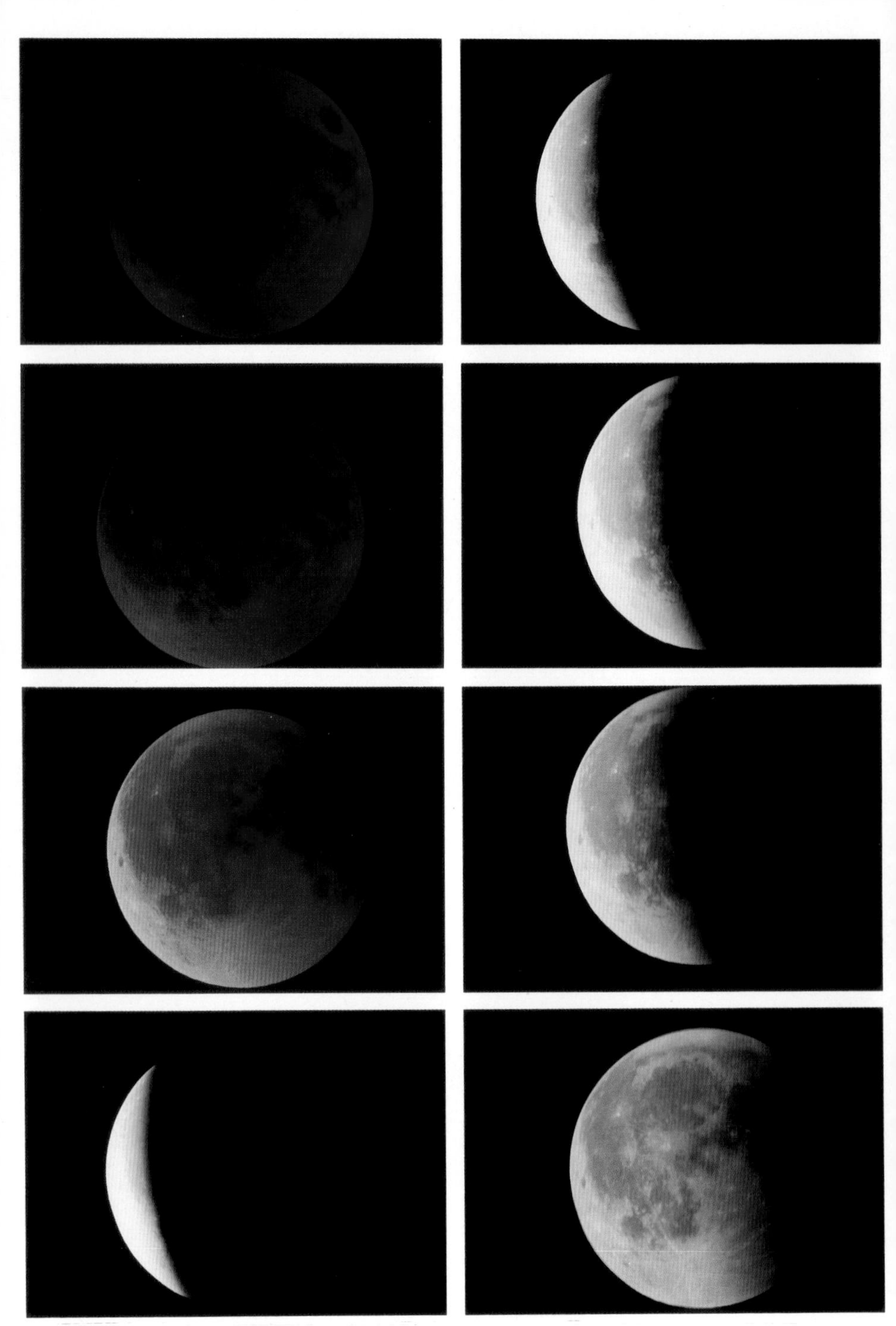

The total lunar eclipse of August 16/17, 1989. After totality (above left), the edge of Earth's inner shadow (umbra) can be seen traveling across the lunar surface.

Astronomical Calendar 2001–2010

2001

January begins with Venus shining brightly in the west, in Aquarius, as the evening star. As it gets dark, Jupiter and Saturn become visible in Taurus high in the east. By 9:30 P.M. or so, from mid-northern latitudes, these latter two are almost straight overhead while Venus has set.

The Moon appears close to Saturn on January 5 and just east of Jupiter on January 6. A total lunar eclipse takes place January 9, visible mainly from Africa, Europe, and Asia.

Mars rises in the southeast at 3 A.M. at the beginning of the month and at 2 A.M. at the end of the month. Mercury may be seen briefly in the evening during the last week of January.

In **February**, Venus completes its rise in the west and begins to descend later in the month toward the horizon. Jupiter and Saturn remain in Taurus. Mars still cannot be seen until 1 A.M., even at the end of February. Mercury appears in the sky shortly before sunrise.

Venus continues to shine brightly as the evening "star" during the first two weeks of **March**. Later in the month, it disappears into the evening twilight. Jupiter and Saturn continue to shine in Taurus, but are now about halfway up in the western sky, now due to Earth revolving around the Sun. Mercury is well placed for observation in the morning sky for southern observers by mid-month, and Mars is in the south as twilight begins.

Jupiter and Saturn remain in Taurus as **April** begins, continuing to sink lower into twilight as the month progresses. By the end of April, Jupiter, which is always much brighter than Saturn, is still easy to observe, whereas the more distant planet will be a real challenge or impossible to see, setting only about an hour after the Sun. Mars officially enters the evening sky, rising just before midnight at the end of the month. Venus has now become the morning "star," rising in the east before sunrise. It moves much higher in the sky, right now for Southern Hemisphere observers.

On **May** 15, Mercury, which has moved quickly into the evening sky during May, passes Jupiter. Both become lost in twilight for a while as the month comes to an end. Mars is now rising at 10:30 P.M. and Venus is well up in the east at sunrise by mid-month.

During **June**, the only planet visible in the evening is Mars. At opposition on June 14, it is now rising at 9 P.M. Venus is still in a good position in the morning sky before sunrise and Saturn joins it low in the east at 4 A.M. A total solar eclipse occurs on June 21, visible from South America, the Atlantic Ocean, and South Africa.

Mars continues to be the only evening planet during **July**. Southern observers can spot Jupiter and Mercury appearing near each other in eastern Taurus as both emerge from the twilight as the month begins. A partial lunar eclipse takes place on July 5, visible from South Africa, Asia, Australia, the Pacific Ocean, Mexico, and the extreme western United States.

Later, Jupiter becomes more visible as Mercury disappears in the twilight glare. Higher, Venus and Saturn "meet" mid-month, as Venus moves toward the horizon each night heading toward Jupiter as **August** approaches (see diagram 1, below).

A conjunction of Jupiter and Venus occurs on August 6 in the morning sky (see diagram 2, below). This is followed by Venus disappearing by month's end, as it heads toward the evening sky again. Saturn

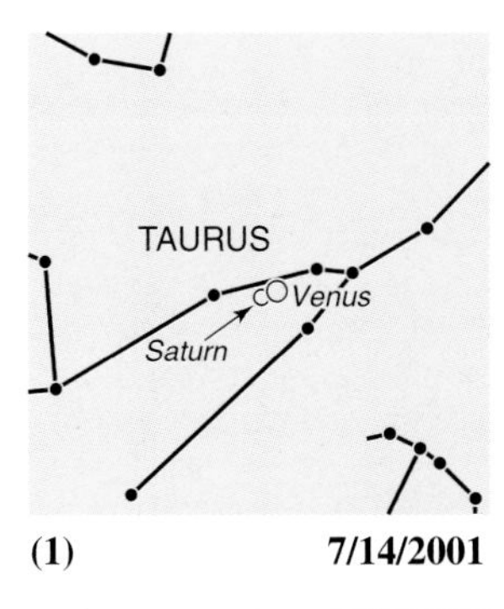

(1) **7/14/2001**

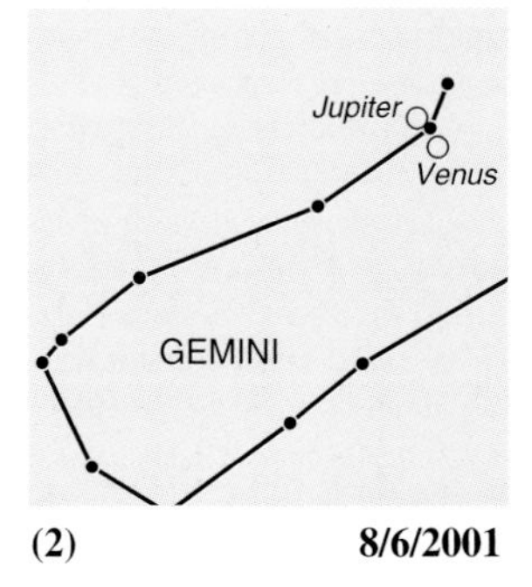

(2) **8/6/2001**

remains in Taurus, rising at 2 A.M., leaving Mars, which sets as Saturn rises, alone again this month in the evening sky.

Mercury reappears in the evening sky for southern observers. Mars is in the south at sunset and remains in the sky until after midnight. Saturn, in Taurus, appears in the northeast at about 1 A.M., with Jupiter following at 2 A.M. in Gemini. Venus rises about an hour and a half before sunrise at 4:30 A.M.

Mercury remains well placed for Southern Hemisphere observers throughout most of **September** and Mars, in retrograde, stays almost due south throughout the month at sunset, although it moves from Scorpius into Sagittarius during the month. Saturn enters the evening sky, rising at midnight by mid-September. Jupiter appears in the east by 2 A.M. and Venus lingers in the predawn sky until late in **October**.

On October 27, Venus and Mercury appear very close together while Jupiter has officially become an evening planet again, rising at midnight by mid-month. Saturn still precedes it in Taurus, and Mars continues to retrograde from Sagittarius to Capricornus.

November brings Mars out of Capricornus into Aquarius, and Saturn into the sky very shortly after sunset, giving observers two planets to look at after it first gets dark. Jupiter joins them at 8:30 P.M. In the morning sky, both Venus and Mercury move too low to be observed easily, seeming to remain very close in the sky as they go.

The apparent distance between Mars and Saturn decreases very rapidly during **December**. Saturn reaches opposition on December 4. An annular eclipse of the Sun can be seen December 14, but only from the central Pacific Ocean and a small band in Central America. Partial phases of this eclipse can be seen from the western United States, Mexico, and Central America as well.

Late in the month, Mercury emerges from the evening twilight, and may be observable by the end of the year. There will be four planets in the evening sky at this time, but when Mercury sets, Jupiter may still be too low to be seen.

2002

Mercury reaches a good viewing location for midlatitudes in **January**. Mars, in Pisces, starts the evening high in the west. It is followed across the sky by Saturn in Taurus and Jupiter in Gemini. Jupiter is in opposition on January 1, so it's up almost all night. The other three planets set at about 6:30 P.M., 10 P.M., and 4 A.M.

Mercury leaves the evening sky before **February** begins, and Venus can be found low in the western sky by the end of the month. Mars, Jupiter, and Saturn appear to be getting closer together as Mercury appears in the morning sky for a while until the last week of **March**.

By the beginning of **April**, four planets appear lined up in the western sky at sunset. Starting from the horizon, they are Venus, brightest in Pisces, Mars in Aries, Saturn in Taurus, and Jupiter in Gemini. As the month progresses, Venus, Mars, and even Mercury all come together in Taurus low in the west. Mars seems to pass Saturn on **May** 3; Venus passes it on May 7 (see diagram 3, below); then, Venus overtakes Mars on May 10.

As Saturn and Mars drop from view later in the month, Venus moves into Gemini to meet up with Jupiter by **June** 4 (see diagram 4, below), and finally Jupiter descends into

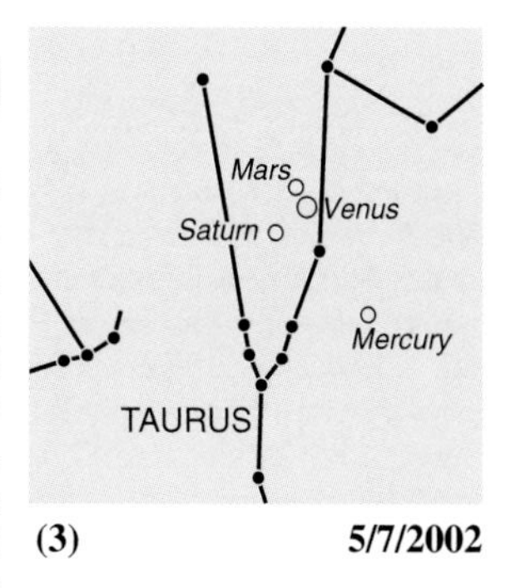

(3) **5/7/2002**

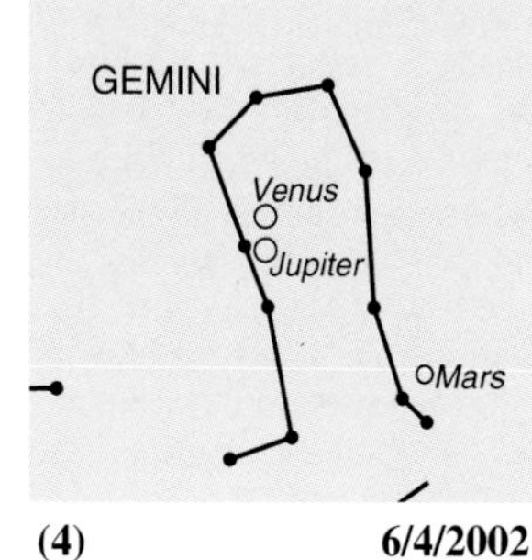

(4) **6/4/2002**

the twilight as July approaches, leaving Venus as the only evening planet. On June 10, another annular eclipse can be seen only in the Pacific Ocean. Partial phases are visible from western North America.

Meanwhile, during the month of June, Mercury reappears in the morning sky to be joined by Saturn late in the month. These two faint planets form a close pairing on **July** 1, then Mercury descends into twilight and is lost from view.

In **August**, Jupiter appears in the morning sky as Venus moves lower, for observers in the evening. For southern observers, it remains higher in the sky throughout August and **September**. Mercury makes a good appearance from mid-August through mid-September for the Southern Hemisphere as well.

By the beginning of **October**, Saturn is rising before midnight again, with Jupiter following three hours behind. Mars has become visible again just before dawn, but Venus disappears from the evening sky.

November starts with all five planets up in the morning sky, but Venus and Mercury are too close to the Sun to be seen. Saturn now rises by 9 P.M., as it approaches a December 18 opposition and is up the rest of the night. Jupiter moves into the evening sky again this month, rising at about midnight at the beginning of the month and before 11 P.M. by the end. Mars can be seen from 5 A.M. until lost in twilight. Now, Venus dominates the morning sky, appearing close to Mars the first week of **December** (see diagram 5, right). As the year draws to an end, Venus appears to separate from Mars and get lower each night. A total eclipse of the Sun is visible on December 4 from South Africa, Australia, and the Indian Ocean.

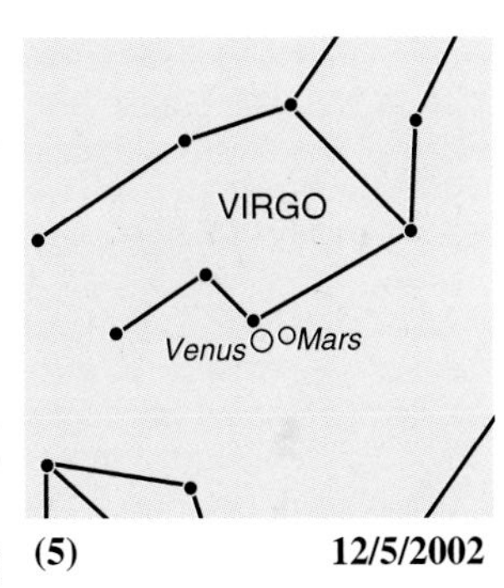

(5) 12/5/2002

2003

Saturn and Jupiter remain easily visible in the evening sky throughout **January**, with both now visible right after dark. In the morning sky, Venus still dominates, but Mars is higher and Mercury makes an appearance late in the month. Things remain pretty much the same during February, March, and April except for the ever-changing trek against the background constellations.

Jupiter reaches opposition at **February** 2, and Mercury moves out of the morning sky again by the beginning of **March**. During that month, Venus, though still far from the Sun, follows the ecliptic to the south, making it more difficult to spot from the Northern Hemisphere. Southern observers will enjoy it as the morning star until the end of May. On **May** 15–16, a total eclipse of the Moon can be observed from North and South America and an annular eclipse from Iceland and the Arctic on May 30.

Mercury remains difficult to see, as it does not appear far from the Sun during this time. The best times to spot it are when it appears just south of Venus in the morning sky of May 26 and even closer on June 20 (see diagram 6, above).

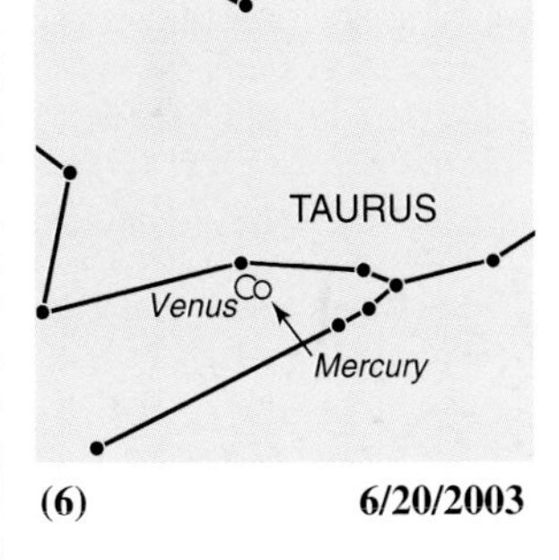

(6) 6/20/2003

During **June**, Saturn drops too low in the west to be seen after sunset leaving Jupiter as the only evening planet. In fact, on **July** 1, Mars is the only other planet that can be seen in the sky. This situation remains until Mercury joins Jupiter for a conjunction on July 25. This event is very difficult to see except from the Southern Hemisphere, where the angle of the ecliptic allows the two planets to appear high enough not to be completely obscured by evening twilight.

As **August** approaches, Mercury moves farther from the Sun, as Jupiter drops quickly into twilight. Mercury can be seen well from the south, but for the rest of the world, Mars is now the only planet visible in the evening. Heading toward opposition on August 29, it rises in the east by 10 P.M. at the beginning of the month.

In the morning sky, Saturn has reappeared, shining faintly in the twilight in Gemini. It is not until late **September** that things begin to change much again as Jupiter enters the morning sky and Venus appears low in the evening sky as seen from the Southern Hemisphere. Saturn enters the evening sky, rising by midnight in early **October**, but Mars remains the dominant planet for northern observers as Venus remains too low to be seen well.

During **November**, Venus finally moves to a more favorable position for the Northern Hemisphere, but remains low in the southwest at sunset. Mercury is there, too, but is very difficult to see. A lunar eclipse is visible on the night of November 8 from eastern North America, Europe, and West Africa. A total solar eclipse takes place on November 23, but it can only be seen from Antarctica.

In **December**, Jupiter rises before midnight in Leo. With Saturn in Gemini and Mars setting in the west in Pisces, the year closes with these three bright evening planets in the sky.

2004

In **January**, Saturn is at opposition on New Year's Day. Venus moves upward in the west so northern observers can see it better. Mars is high in the south at sunset, with Saturn rising up in the east. Jupiter follows later at night, rising in Leo around 10:30 P.M. Except for the fact that Jupiter and Saturn are still visible at sunrise, there are no morning planets other than Mercury, which appears briefly again in the Southern Hemisphere.

February brings the four evening planets closer together in the sky, with Venus shining brightly now as it appears to approach Mars. Jupiter reaches opposition on **March** 4 and is, therefore, up all night. Mercury moves into a favorable position in the evening sky mid-March and may be seen in Pisces below Venus for a while. **April** 19 brings a partial solar eclipse, visible from the South Pacific and South Africa. Its companion lunar eclipse occurs on **May** 4, and can be seen from South America, the Atlantic Ocean, Africa, western Asia, and the Indian Ocean.

At the beginning of May, Venus, Mars, and Saturn form a large, flat triangle in Taurus and Gemini, but Venus quickly turns during the month to drop ahead of the others toward the horizon. Mars passes Saturn very low in the west on May 22 (see diagrams 7 and 8, above). Meanwhile, Mercury makes a good appearance in the morning sky, but only for Southern Hemisphere observers. Throughout **June**, Jupiter is the only evening planet that is easy to spot.

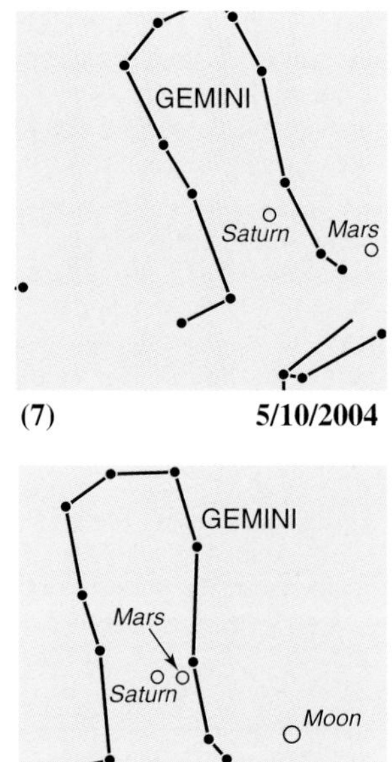

(7) 5/10/2004

(8) 5/21/2004

On **July** 9, Mercury and Mars meet low in the northwest, an event best viewed from southern latitudes. Jupiter is still high in the west even though summer brings longer days and a late sunset in the Northern Hemisphere. Venus, quite well placed for more southerly observers throughout the month, is visible with a little more difficulty in the north as well. Saturn joins Venus in the evening sky by month's end.

August brings Mercury fairly high in the west at sunset

for Southern Hemisphere observers. It appears between Mars, now very low in the evening twilight, and Jupiter, which is higher and brighter. All three planets are in Leo. By the end of the month, Mars and Mercury disappear from the sky as Jupiter slowly approaches the same fate. At the same time, Venus and Saturn have been getting closer together until they pass on August 26 and Venus heads into the solar glare (see diagram 9, right). **October** brings another pair of eclipses. On October 13, a partial solar eclipse can be seen from the North Pacific and Asia and on October 27, a total lunar eclipse can be seen from most of North and all of South America as well as the western portions of Europe and Africa.

Through October, Saturn gets slowly higher each night and Venus gets lower. On **November** 5, Venus and Jupiter make a nice sight in the morning sky, appearing to pass close to one another as they switch places, leaving Jupiter higher and Venus even lower (see diagram 10, right). Late November is another good time for southern observers to spot Mercury in the southwest just after sunset.

Venus and much fainter Mars meet on **December** 5 where they can be seen low in the southeast, as dawn approaches (see diagram 11, right). Finding Mars is made much easier due to its proximity to Venus in the sky at this time. The same assistance in finding Mercury is possible on December 28, when Venus passes its tiny neighbor very low in the southeast. This month, Saturn, rising about 8 P.M. mid-month in Gemini, is the only evening planet.

2005

Mercury and Venus linger near each other very low in the sky during the first week of **January**, with Mars above them in the southeast. All of these planets are best viewed at this time from as far south as possible, but Venus is so bright that it aids in finding the others. Jupiter is high in the south and Saturn is low in the northwest before sunrise. During the month, Jupiter, in Virgo, joins Saturn, which reaches opposition on January 14, in Gemini, as an evening planet, rising about midnight.

February begins with Venus and Mercury too low to be seen, but Mars rising higher in the morning sky. Saturn is up at sunset and Jupiter appears at 11 P.M. These two planets will remain together in the evening sky until early July, when Saturn goes into conjunction with the Sun. Jupiter is at opposition **April** 4. An annular/total eclipse takes place on April 8, visible in the South Pacific and a small part of South America. Partial phases can be seen from the southern United States, Mexico, and Central America.

Early **May** brings Mercury into a good position for observing in the morning sky. Look

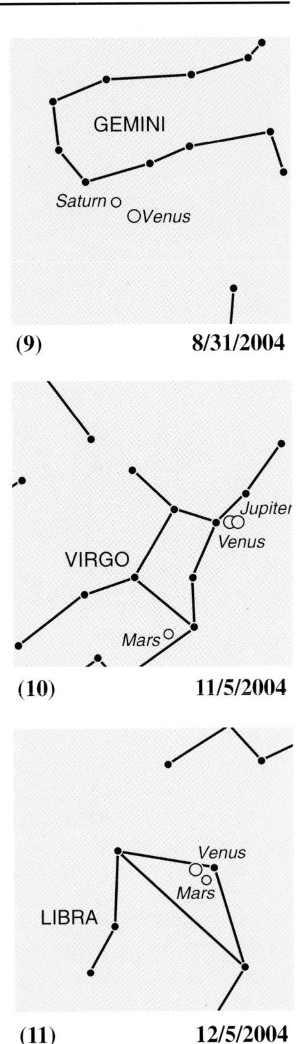

(9) **8/31/2004**

(10) **11/5/2004**

(11) **12/5/2004**

for it below the thin waning crescent Moon on May 5. Later in the month, Venus becomes the evening "star" again, appearing to rapidly approach Saturn low in the northwest. Mercury joins the pair around **June** 24, with a very close pairing of Venus and Mercury

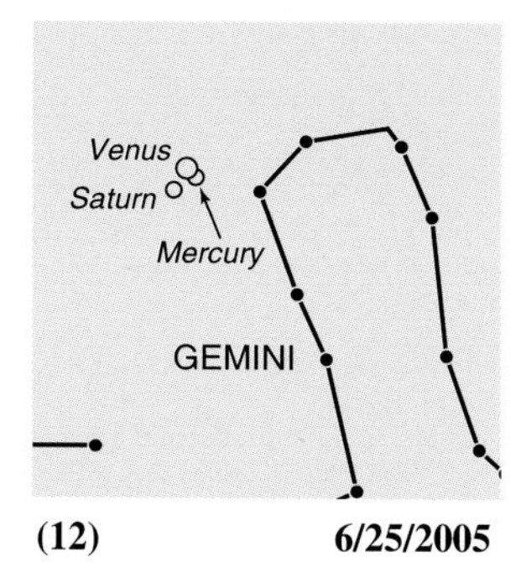

(12) **6/25/2005**

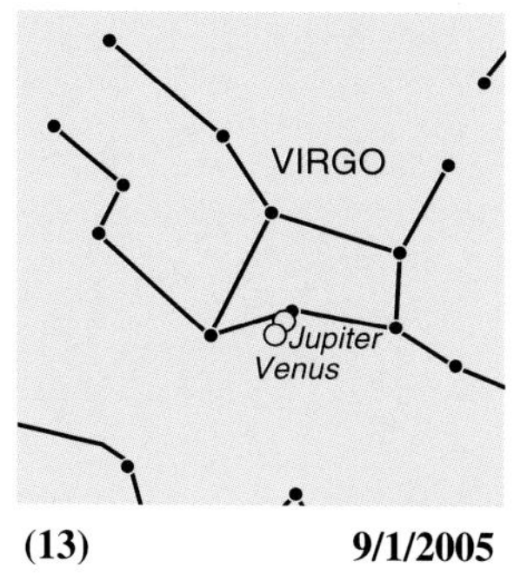

(13) **9/1/2005**

occurring on June 26 (see diagram 12, above).

As **July** begins, Saturn is too low in the west at sunset to be seen easily, but Venus and Mercury are still visible appearing close together until the middle of the month. Jupiter is now high in the southwest at sunset, setting about 10:30 P.M., and getting lower each night while Venus gets higher for a while. Mars becomes an evening planet after the first week of **August**.

September 1 brings a conjunction of Jupiter and Venus in the western sky—always a spectacular occurrence. This takes place in Virgo (see diagram 13, above). Saturn stands out now against the dim stars of Cancer in the morning sky, passing near the Beehive cluster on September 23. Though high enough when seen from the Southern Hemisphere, Jupiter and Venus move lower in the sky as seen from the north.

By **October**, Mars rises at 9:30 P.M. making it the dominant evening planet once Venus has set. Saturn rises about five hours later and the two are up the rest of the night. Two eclipses occur this month. The annular eclipse of the Sun on October 3 can be seen from Spain, across central Africa and the Indian Ocean. Partial phase visibility extends throughout Africa, Europe, the Middle East, and India. On October 17, a partial lunar eclipse is visible from North America, the Pacific Ocean, Australia, and Asia.

Venus continues to be visible (best from southern latitudes) throughout **November** and **December**. Mars, at opposition on November 7, is also up in the east by the time it gets dark. Saturn becomes an evening planet, up around midnight at the beginning of November and by 8 P.M. by the end of the year.

2006

Venus can still be seen low in the west at sunset for the first several days of **January**. It reappears in the morning sky late in the month. Mars is high in the east and is joined by Saturn after 8 P.M. Jupiter rises about 3 A.M. at the beginning of the month and earlier as the month progresses. On January 28, Saturn reaches opposition to the Sun bringing it into the sky at sunset during **February**. The Moon passes close to Mars on February 5. Venus becomes prominent in the morning sky by the end of the month.

The highlight for **March** is an especially good appearance of Mercury for southern observers and a total solar eclipse on March 29. This can be seen from West Africa and Asia with partial phases visible throughout most of Africa, all of Europe, and much of Asia. Mars, Saturn, Jupiter, and Venus shift against the constellations throughout the month, with the only thing of note being that Jupiter enters the evening sky this month. It reaches opposition **May** 5.

Not much changes through **June** 14, after which Mars appears to pass near Saturn. Both planets are also near the Beehive star cluster in Cancer, making a pretty sight through binoculars (see diagrams 14, 15, and 16 on page 125). At this same time, Mercury moves through Gemini in one of its best appearances this decade for Northern Hemisphere observers. Venus continues all this time as the morning "star," rising over an hour before the Sun even at the beginning of **July**.

As Saturn continues to sink into twilight each evening, Mars gets lower, too, but can still be seen in Leo through the end of July.

Mercury joins Venus in the morning during **August**, appearing close to it the first

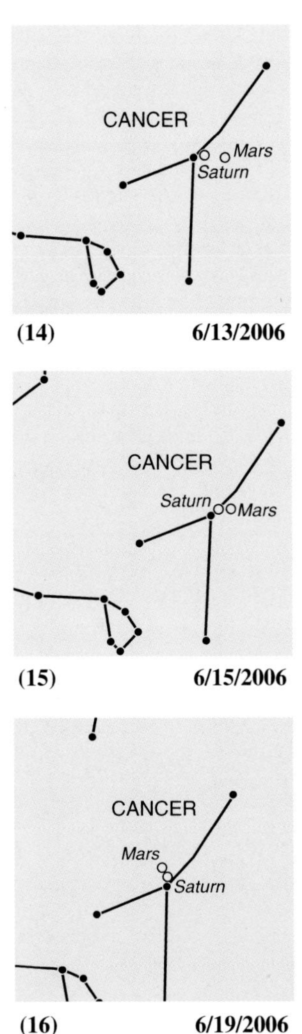

(14) **6/13/2006**

(15) **6/15/2006**

(16) **6/19/2006**

week of the month. Both pass Saturn on their way back into the solar glare, Mercury on August 21 and Venus on August 26. The conjunction of Venus and Saturn will be very low in the sky, but worth looking for as the planets will be only 7/100ths of a degree apart.

By the end of August, Mars is so low that it is difficult or impossible to see. This leaves Jupiter to dominate the evening sky until early November. Before the largest planet disappears, however, southern observers will see the second smallest planet pass by it twice in the sky at the beginning of the month.

September brings two eclipses. A partial lunar eclipse visible from Africa, Europe, Asia, and the Indian and Atlantic Oceans. The solar eclipse of September 22 can be seen northeast of South America and from the Atlantic Ocean.

Replacing Jupiter in the evening sky, Saturn rises at 11 P.M. at the beginning of **December**. Not until near the end of the month does Venus appear in the evening sky just after sunset and Jupiter then Mars appear in the morning sky.

2007

During **January** and **February**, Venus emerges in the evening sky, to the west, joining Saturn, which shines in Leo now at sunset. In the morning sky, Jupiter and Mars rise earlier each night. Saturn is in opposition on February 11.

March brings Jupiter into the evening sky and a good appearance of Mercury just before sunrise, especially for the Southern Hemisphere. This apparition continues well into **April**. On March 3, a total eclipse of the Moon is visible from Africa, Europe, and western Asia, with partial phases seen from North and South America and the rest of Asia. On March 18, a partial solar eclipse can be seen from Asia and western Alaska.

During **May**, Venus, Saturn, and Jupiter can all be seen in the evening sky, first only in theory, because Venus (in the west) and Jupiter (which reaches opposition in the east on June 6) are quite low as the month begins. By the end of May, however, they both get higher and even Mercury enters the picture by **June** 1.

Although Mercury never catches up with Venus, Saturn seems to come down to meet the second planet in a conjunction on June 30 (see diagram 17 on page 126). During **July**, these planets seem to separate as they plunge into the twilight to disappear by **August**. Near the end of August, a total lunar eclipse occurs. The morning of August 28, it can be seen as partial from North and South America. Totality is visible slightly from the far western states.

Jupiter remains the only evening planet until Mercury makes an appearance in early **September**. Mid-month, Mars begins rising before midnight. In the morning sky, Venus and Saturn have moved far enough from the Sun to be seen again. September 11 also brings a partial solar eclipse, but it is only visible from southern South America, the southern Atlantic Ocean, and Antarctica.

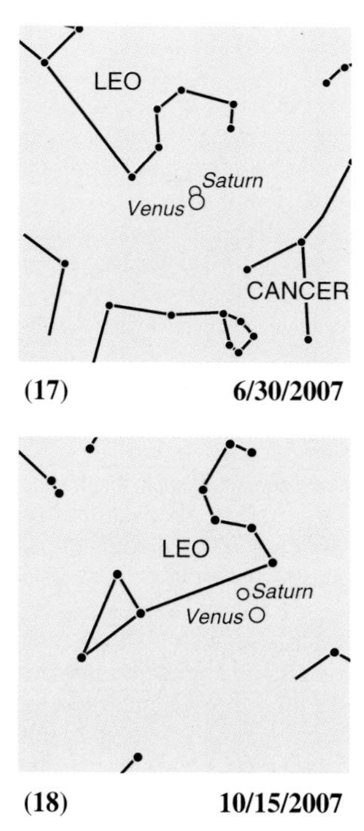

(17) **6/30/2007**

(18) **10/15/2007**

By **October**, Mercury has sunk too low for northern observers to see, but is still well placed as seen from the Southern Hemisphere. Jupiter is high in the south at sunset and Mars rises in the late evening in the northwest. As it heads back into twilight, Venus passes Saturn on October 17 (see diagram 18, above).

Jupiter finally leaves the evening sky as **December** begins, so for a while, until Mars rises about an hour after sunset, there are no bright planets in the early evening sky. However, Mars reaches opposition on Christmas Day and Saturn has begun rising before midnight. Later, two planets, then three, can be seen after Venus rises at about 4:30 A.M.

2008

Mars is visible right after sunset during **January** in the northeast. Saturn rises around 10 P.M. and Venus at 5 A.M. On **February** 1, Venus and Jupiter appear fairly close together in Sagittarius, and then Mercury moves in to meet up with Venus in Capricornus on February 25. This latter event is best viewed from the Southern Hemisphere (see diagrams 19 and 20, right). February 6 brings an annular solar eclipse, but an unfavorable one. The night of February 20 a total lunar eclipse is visible from North and South America, most of Europe, and western Africa. February 25 brings Saturn to opposition with the Sun.

During **March**, Mars shines brightly almost straight overhead at sunset from mid-northern latitudes, with Saturn high in the eastern sky. The two seem to be getting closer together while they both shift westward. By the end of **April**, Venus is lost in the morning twilight.

Jupiter enters the evening sky by the end of **May**. On **July** 9, Mars appears to pass Saturn and Jupiter reaches opposition. Later in the month, Venus appears in the evening sky while Jupiter has begun rising before sunset. This brings all but Mercury into the early evening sky.

On **August** 1, another unfavorable eclipse, this time a total eclipse, occurs. Totality is only visible across extreme northern Canada and Greenland as well as Siberia and Russia. Partial phases can be seen from much of Europe, however. On August 16, there is a partial lunar eclipse that can be seen from South America, Africa, Europe, and Asia. In mid-August, Venus, then Mercury, pass Saturn very low in the western sky, though this must be viewed with a very flat horizon and as soon after sunset as possible. All three planets have set by $1^1/_2$ hours after sunset (see diagrams 21, 22, and 23 on page 127).

Mars, Mercury, and Venus form a triangle on **September** 8, although this event will be

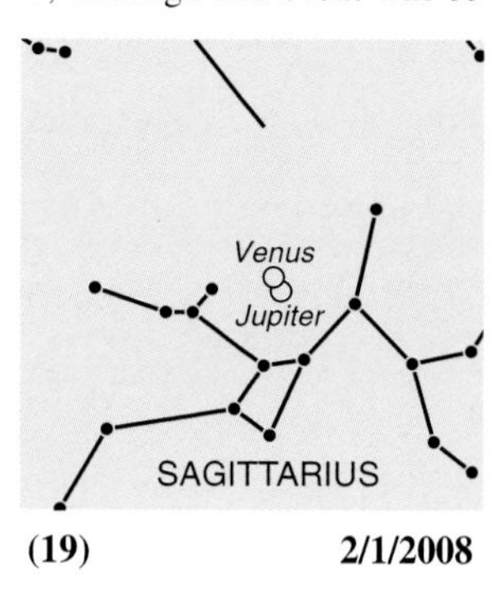

(19) **2/1/2008**

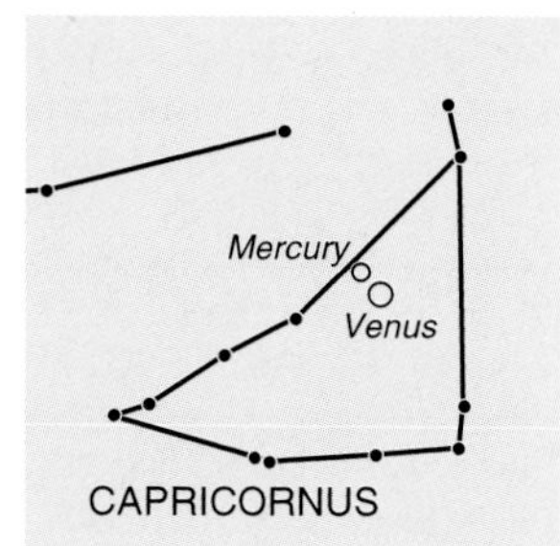

(20) **2/25/2008**

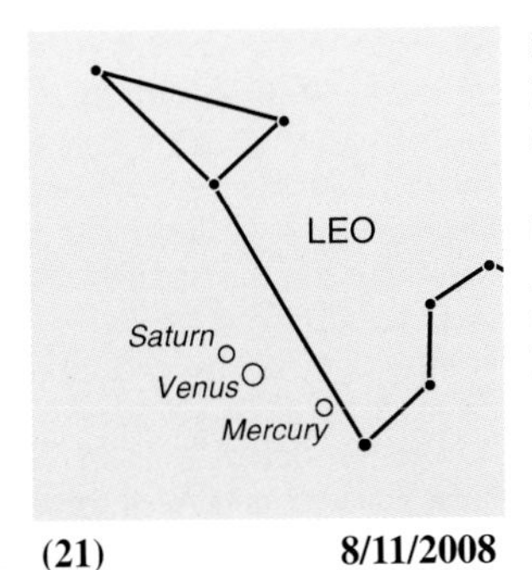

(21) 8/11/2008

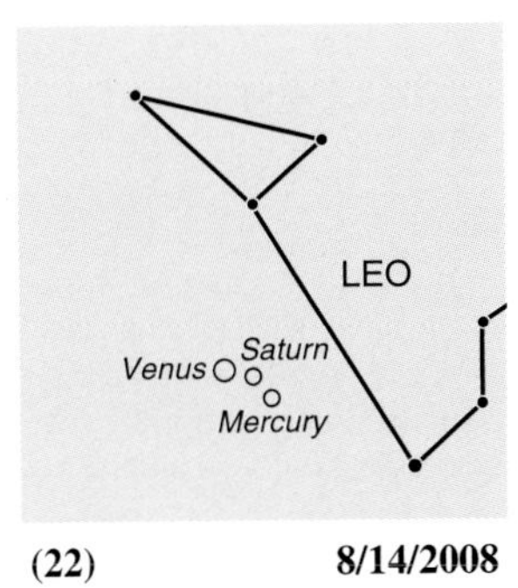

(22) 8/14/2008

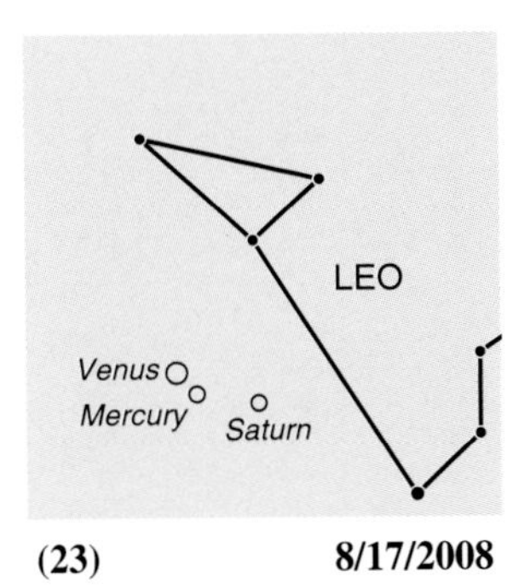

(23) 8/17/2008

very difficult to see from all but the southern latitudes. Jupiter is now high in the south by 9 P.M., making it the dominant evening planet for the rest of the year, but Venus still lingers for a while low in the southwest after sunset. By the end of September, Saturn reemerges from the morning twilight to begin another cycle around the sky. It is now in Leo.

Mars and Mercury respectively disappear in the solar glare by early and late **October**. **November** and **December** bring only a close apparent approach of Venus to Jupiter on December 1 (see diagram 24 on page 128) and then Mercury very low in the southwest on December 30.

2009

The last view of Jupiter in the evening sky for a while may be glimpsed during the first week of **January**. Venus remains as the evening "star" and the only evening planet until around 10 P.M. On January 26, an annular eclipse of the Sun can be seen mainly over the South Atlantic and the Indian Ocean.

Southern observers will see a close pairing of Mars and Jupiter, with Mercury above them on **February** 17. Mercury drops out of sight as Mars and Jupiter become more visible worldwide during **March**. Saturn reaches opposition March 9. In the meantime, Venus moves out of the evening sky to reappear in the morning in early **April**. It passes Mars on April 23, and remains west of it through **May**. It is as if Venus and Mercury have switched places now as Mercury moves into the best appearance of the year in the evening sky.

After Mercury drops back toward the horizon in mid-May, Saturn remains the sole evening planet until nearly the end of **June**. At this time, Jupiter begins to rise before midnight. On June 19, Venus again passes Mars, moving to the east of it as it heads back to conjunction with the Sun. Early in the month, Mercury can be seen below them for a while.

July 21 brings a total solar eclipse to Asia and the Pacific Ocean areas. Jupiter reaches opposition August 15 and Mercury passes Saturn on August 18, but both planets are quite low in the sky. Because Mercury is so small and Saturn is so far away at this time, this conjunction will be difficult, though not impossible for northern observers to see. As with many apparitions of Mercury, the farther south the observer, the better.

By mid-**September**, Mercury and Saturn leave the evening sky, so Jupiter is the sole evening planet for a while. This changes when Mars begins rising before midnight in **November**. Before this time, Venus passes Saturn **October** 13 as the fainter planet emerges from the morning twilight glow and the brighter one drops into it.

In mid-**December**, Mercury is again well placed for southern observers, appearing low in the southwest at sunset. Saturn begins to rise by midnight at the end of the month. The year ends with a partial lunar eclipse on December 31. It is visible from the Atlantic Ocean, Africa, Europe, Asia, and the Indian Ocean.

2010

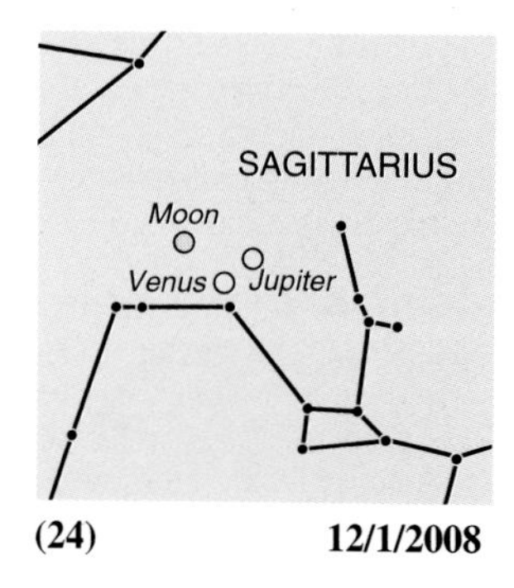

(24) **12/1/2008**

2010

The final year of the decade begins with Jupiter high in the southwest at sunset in Aquarius. By 9 P.M., Mars (at opposition on **January** 30) rises in Leo, with Saturn in Virgo just three hours behind. Mercury is still in the evening sky, but too close to the Sun to be seen. An annular eclipse is seen across Africa and the Indian Ocean on January 15 and a partial lunar eclipse from North and South America, Eastern Asia, the Pacific Ocean, and Australia. Late in January, Mercury moves quickly into the morning sky for an excellent appearance for all latitudes.

In **February**, Venus begins to move into the evening sky, passing Jupiter low in the west on February 16. This is only visible because both planets can shine so brightly. Mars is now in the sky at sunset and Saturn rises by 9 P.M. Mercury, always very quick to move, and Jupiter, much slower, both descend out of view by **March** 1. On March 22, Saturn reaches opposition.

Venus, moving ever higher is approached by Mercury on **April** 6, but the innermost planet quickly moves again into the glare of twilight. During **May**, Venus, Mars, and Saturn form a shrinking string of lights along the ecliptic. Note the different colors. Venus is bright white, Mars a dull orange, and Saturn a bit off-white. The trio forms a beautiful triangle in the western sky the week of **August** 5 (see diagram 25, right). For Southern Hemisphere observers, Mercury also enters the picture several degrees below the conjunction. For all, the thin crescent Moon will add to the scene on August 12. **July** 11 brings the final solar eclipse of the decade. Totality is visible only in the Pacific Ocean and South America.

While all of this has been going on, Jupiter, which is officially in the evening sky, does not rise early enough to be in the sky with the other planets. Only in late **September** does Jupiter, at opposition on September 22, rise early enough (about 7:30 P.M.) that Venus and Mars are still in the sky when it can be seen. During September, while Mars and Venus remain in the same part of the sky, Saturn seems to have moved off and is getting lost in the evening twilight.

Venus, then Mars, suffer the same fate, but first Southern Hemisphere observers are treated to a double conjunction of Mercury and Mars during a twenty-day period in **November** and **December**. As Mars gets lower in the sky, Mercury moves upward, passing Mars on November 21, then rising above it and coming back on December 12. Watch for the thin crescent Moon near Mars on December 6. This will guide you to sighting this difficult conjunction of two dim planets.

On December 21, a total lunar eclipse can be seen from all of North America and much of the Pacific, with partial phases visible from South America, Europe, and western Africa.

The decade ends with Jupiter, once again, the sole evening planet, but Saturn is high in the south before sunrise and Venus makes a fine morning star, rising more than three hours before the Sun.

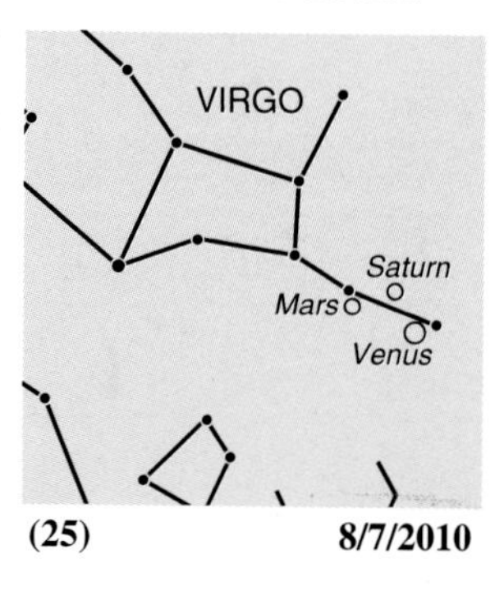

(25) **8/7/2010**

Star tracks around the south celestial pole. ▷

Lexicon of Celestial Bodies

Achernar
Achernar is the ninth brightest star in the sky. Its name comes from the Arabic expression Al Ahir al Nahr (the end of the river), which was later corrupted into the present form. The name refers to the star's position in the constellation Eridanus, a long drawn out string of stars winding across the sky and terminating in the south with Achernar. Achernar can be seen only south of 33 degrees northern latitude and is thus a star primarily of the Southern Hemisphere. Achernar is about 88 light-years away from Earth and shines 650 times more brightly than the Sun. Its diameter is estimated to be seven times that of the Sun.

The globular star cluster 47 Tucanae.

Algol
Algol is the second brightest star in the constellation Perseus and also the most famous variable star (see pages 150–51). Its brightness fluctuates so much that the change can be noticed even without binoculars. This unusual behavior seems to be reflected in the star's name, which is derived from the Arabic expression Al Ra's al Ghul, which means demon's head. But despite this apparent causal connection there is no indication that astronomers of ancient times, whether Greek, Roman, or Arab, were aware of the phenomenon of variable stars.

In old depictions of the constellation Perseus (see pages 30–31) Medusa's head is located where the demon star is. Algol consists of two stars that revolve around each other and alternately cover each other up (as in a solar eclipse) in a regular cycle, causing the light of Algol to dim.

Today we know that Algol is about 100 light-years distant from Earth and that its two stars revolve around each other in exactly 2 days, 20 hours, and 49 minutes. When the darker star moves in front of the brighter one, the luminosity of the binary declines by a factor of about 3 for 10 hours. This dimming is clearly visible. The brighter of the two stars is 100 times brighter than the Sun and has a diameter of 2.8 million miles (4.5 million km). It is thus one of the so-called supergiant stars. Very little is known about its darker companion star. Because it covers up the brighter star for such a long time span it must be even bigger, but at the same time it is considerably less bright and probably has less mass as well.

Alpha Centauri
This is the only one of the ten brightest stars that generally is referred to by its scientific designation—first introduced by Johann Bayer—rather than by a common name. In 1603, Bayer devised a system of assigning letters of the Greek alphabet to the stars, followed by the genitive form of the constellation's Latin name. The stars within a constellation are listed in order of apparent brightness. Alpha is the brightest star in a constellation, Beta the second brightest, Gamma the third brightest, and so on. Thus, Alpha Centauri is the brightest star in the constellation Centaurus (see pages 86–87). This unusual star also has proper names, such as Rigel Kentaurus, foot of the centaur, or Toliman, but they have never gained wide usage.

Today Alpha Centauri is one of the best known of all the stars because it is, along with its companions, our closest stellar neighbor. It is a mere 4.3 light-years away, which amounts to 25 trillion miles (43 trillion km), a distance light travels in 4 years and 4 months. Alpha Centauri is thus 275,000 times farther away from Earth than the Sun. The Scottish astronomer Thomas Henderson first measured the distance of this star in 1839 from the observatory in Cape Town. But 150 years earlier the missionary Richaud had realized when examining Alpha Centauri in India through a small telescope that this was a binary star. It has in the meantime become one of the best known and most closely observed double stars. We now know that the two stars (Alpha Centauri A and Alpha Centauri B) revolve around each other once every 80 years and that the distance that separates them fluctuates between 1 and 3.2 billion miles (1.6 and 5.2 billion km). Both stars have approximately the same mass as the Sun, and the brighter of the two, Alpha Centauri A, resembles the Sun closely in other respects as well.

In 1915, the astronomer Innes discovered another, fainter star in the immediate vicinity of Alpha Centauri, a small star that, when measured in greater detail turned out to be even a little closer to Earth. It is called Proxima Centauri, the closest one in Centaurus, and is exactly .065 light-years or 382 billion miles (615 billion km) closer to us than Alpha Centauri. Proxima Centauri can, of course, be seen only through very large telescopes. Measuring 43,000 miles (70,000 km) across, it is 2,000 times less luminous than the Sun. It has not been settled conclusively to this time whether Proxima Centauri is physically part of the binary Alpha Centauri and thus bound to the larger stars by gravity or whether it appears in the same line of vision only by chance.

Arcturus

The watcher of the bear, as its name translates, belongs to the constellation Boötes (the Herdsman). Arcturus is the brightest star in the northern half of the celestial sphere, and many people think that it was the first star to have been given a name. Arcturus is 34 light-years distant and thus in the immediate neighborhood of our solar system. Its luminosity is about 110 times greater than that of the Sun, and the hot gases of which it is composed are 3,000 times more rarefied than those of the Sun. Arcturus is a red giant, a type of star that has expanded to an extreme degree. It is "cool" (surface temperature: about 4,200° C), and its gases are more diffused than gases in an airless space on Earth, that is, in a vacuum. It is only because of the huge size of these stars that enough energy is emitted from their surfaces to make them visible in spite of their astronomical distances from Earth.

In autumn of 1933, Arcturus had a moment of "world" fame when its light, conducted to a photoelectric cell, opened the World's Fair in Chicago. The light ray activated a switch that turned on the floodlights on the fair grounds. Arcturus was chosen for this because its distance was then estimated to be 40 light-years. Consequently the light ray that triggered the ceremonious opening presumably would have been generated and sent from Arcturus in 1893, when the World's Fair had last been held in Chicago.

Betelgeuse

Orion, one of the most impressive constellations in the sky, is also the only one to include two of the sky's brightest stars, namely Rigel (see page 146) and Betelgeuse. The names of both stars refer directly to the figure of Orion, the Hunter (see pages 54–55). Betelgeuse is a much corrupted form of the Arabic phrase meaning armpit of the giant, for Betelgeuse is shown in old representations in one of Orion's shoulders.

Betelgeuse is easy to spot even without binoculars because of its reddish color. It is a type of star that is called red supergiant by astronomers on account of the truly astronomical dimensions that characterize these stars. Betelgeuse may be as much as 500 to 800 times bigger than the Sun, that is, 430 to 680 million miles (700 to 1,100 million km) in diameter. The considerable difference between these two numbers is not the result of an error in calculation. Betelgeuse is a variable star, and its luminosity undergoes irregular fluctuations. The reason for this is that the star's diameter keeps expanding and contracting. If Betelgeuse were located where the Sun is, it would, when fully contracted, extend as far as Mars's orbit; at greatest expansion it would reach as far as Jupiter. Betelgeuse is approximately 1,040 light-years away from us and is one of the largest stars known.

Binaries

Many "fixed" stars do not travel through space in solitude but have one or more companions. They form double, triple, and multiple star systems, in which the individual partners are held together by mutual gravitational force.

Most binaries (or double stars) reveal their multiple structure only when viewed through binoculars or, more often, through a large telescope with high powers of magnification. Some double stars are so close to each other that their separate bodies cannot be detected at all by visual observation. But precise inspection of their spectrum—the different colors into which their light can be separated—will reveal their double nature. The spectrum changes in characteristic ways when a star moves toward or away from Earth. Stars whose double structure becomes apparent only through an analysis of the spectrum are called "spectroscopic" binaries in contrast to "visual" binaries, whose component stars can be seen with binoculars or a telescope.

Binaries are of great importance in astronomy because they offer the only way of determining the mass of stars with some accuracy. The movement of the stars around each other is subject to the laws of gravity, and the force of gravity depends on the mass of the individual stars and on their distance from each other. The orbital paths of the stars making up a binary, paths that are precisely defined, therefore indicate the effect exerted by the mass and consequently the mass of the two stars. However, plotting the orbital paths of binaries is difficult because the movement along the orbital path is generally very slow. Some binaries take as much as 100 years or more to complete one orbit. Spectroscopic binaries, on the other hand, revolve around each other in as little as 100 days.

The following binaries are marked in the star maps:

- **Mizar-Alcor** in Ursa Major. Anyone with good eyesight can detect these as two separate stars (see pages 32–33). However, it is still unclear whether they actually belong together, that is, whether they revolve around a common gravitational center. The distance between them is ¼ of a light-year, which amounts to 1.5 trillion miles (2.5 trillion km).

Mizar, the brighter of the two, is a visual binary—the first to have been recognized as such. The discovery goes back to 1650 when it was made by the Italian astronomer Riccioli. Mizar is also one of the most famous binaries. It is 88 light-years away. Unfortunately, a small telescope is required to see it as a double star. The two Mizar stars take several thousand years to revolve once around each other. Both of them are in turn double stars—spectroscopic binaries. They are the first of their kind to have been discovered. The American astronomer E. C. Pickering first described them in 1889.

The surface of the planet Mercury—a composite of photographs taken by the Mariner.

• **Albireo** in Cygnus. It often is called the most beautiful of all the binaries. Its double structure is visible through binoculars. The two stars are 410 light-years away and are two completely different colors; one is orange, the other, blue. The stars are 760 and 120 times more luminous than the Sun. The brighter one, which is named Albireo A, is in turn a double star, a spectroscopic binary.

• **Epsilon Lyrae** Like Mizar-Alcor, Epsilon Lyrae can be detected with the naked eye only by people with excellent eyesight. The two components also are referred to as Epsilon 1 and Epsilon 2. When examined through an average size telescope, the two stars turn out to be binaries themselves. The stars of Epsilon Lyrae 1 revolve around each other once in about 1,200 years, those of Epsilon Lyrae 2, in about 600 years. All four stars are of equal brightness and are 150 light-years away from Earth.

Canopus

Canopus, the second brightest star of the sky, belongs to the constellation Carina and is one of the few stars named after a human figure, though not a historical personage but rather a legendary figure from Greek mythology. Canopus was chief pilot of the Greek fleet under Menelaos's command and guided it on the way home after the fall of Troy. He reached the Egyptian coast, where he died. King Menelaos erected a mausoleum in the helmsman's honor and also named the brilliant star that stood just barely above the horizon in the south after him. From the modern point of view, the name could not have been better chosen. Of the ten brightest stars, Canopus is the farthest away from the ecliptic. Spacecraft launched to explore the planets of our solar system therefore use it as a fixed navigational reference point in space by which to orient their flights. Small sensors are trained on Canopus and emit steering signals when it moves out of the sensors' field of vision. All the other bright stars are too close to the ecliptic, on whose plane spacecraft also travel, and the angles between them and Earth and the Sun—the other two points of reference in space—are too small and unreliable to serve as orientation aids.

On the other hand, the extreme southerly position of Canopus has prevented it from being observed much. All the data about it is therefore less definite than than about the other bright stars. Its distance from Earth is given variously at anywhere between 100 and 1,200 light-years. Probably the actual distance is somewhere around 250 light-years. It is quite certain that Canopus is a very large, brilliant star, about 1,400 times brighter than the Sun and 30 times its size. Like the Sun, it seems to be a solitary star; no evidence of a companion star thus far has been detected.

Capella

The She-Goat in Auriga, the Charioteer's shoulder (see pages 48–49), is the only one of the ten brightest stars that never sets for observers in the higher latitudes of the Northern Hemisphere and is thus circumpolar (see pages 9–10). Its distance from Earth is 41 light-years; its mass is three times that of the Sun, and its diameter is 16 times as large as the Sun's.

Capella is a spectroscopic binary (see pages 131–33). What actually is seen when one looks at Capella is two stars that revolve around each other once every 104 days. The mass and diameter of Capella therefore refer to the brighter of the two stars. Capella also has two considerably smaller companions, classified as dwarfs, that can be seen through a telescope. Thus Capella is actually a quadruple star.

However, to date, observation of the four stars has not yet yielded a clear composite picture of this star complex. The invisible Capella twin in particular cannot be measured exactly, and Capella therefore remains partially shrouded in mystery.

Comets

Comets are one of the most striking phenomena in the sky. Large comets with their tails streaming behind them sometimes streak across the entire sky and occasionally can be seen even during daylight. Most comets are invisible to the unaided eye. Every year an average of 20 to 30 comets appear, but most of them can be seen only through a telescope.

Comets revolve around the Sun just as planets do, but they follow very elongated, elliptical orbital paths (see pages 16–17). The time it takes to complete one orbit around the Sun varies from a few years to millions of years. Comets are visible only during a very small segment of their journey, namely when they are in the immediate vicinity of the Sun. There the Sun's radiation heats them up, causing some of their material to stream out from the head and form a tail. This tail is made up partially of dust and partially of gases, which are made to glow by the sunlight. The formation of the tail is strongly influenced by the Sun, which is shown by the fact that the tails of comets always point in the direction exactly opposite from the Sun.

Today comets are pictured as something resembling a dirty snowball. They are not solid bodies like the planets but instead are made up of frozen gases and dust particles lodged in the gases. In their frozen state, comets cannot be seen from Earth at all because they are too small—usually only about 6 miles (10 km) in diameter—and don't reflect enough sunlight. It is only the evaporation caused by the Sun's heat that makes them visible.

Only very few comets have been observed often enough thus far for their orbit and consequently their reappearance in the sky to be calculated. The most famous of these so-called periodic comets is Halley's Comet, which has been observed repeatedly as a brilliant object in the sky over the past 2,000 years. Halley's Comet takes 76 years to complete its orbit around the Sun and will not appear again until the year 2061. The great majority of comets, however, approach the Sun unexpectedly from the depths of the solar system, so that their appearances cannot be calculated ahead of time. Because in their approach to the Sun all comets lose some of their matter in the form of their tails, comets cannot exist indefinitely. The life expectancy of Halley's Comet, for example, is put at only about 200,000 years. The remains of comets continue to travel through space along the same paths and some give rise to meteor showers (see page 139).

Fixed Stars

In contrast to the planets (from the Greek word *planasthai*, wanderer) the stars seem to be motionless, or fixed to their spots on the celestial sphere. But it only looks that way because they are so very far away from us—distances that are measured in light-years (see Fixed Stars Are Not as Immobile as They Seem on page 10). In reality the fixed stars also move in the sky, but they do so at such a slow pace that their motion (called "proper" motion) becomes visually apparent only over thousands of years (see pages 10–11) and otherwise can be proven to exist only through precision measurements.

The fixed stars shine with their own light. They are suns like our Sun, that is, spheres of incandescent gases (called plasma by scientists). In the interior of these spheres, energy is produced through thermonuclear reactions. The number of suns in the universe is vast beyond guessing. The unaided eye can detect about 4,000 stars on a very clear night; a telescope reveals many millions in the region of the Milky Way alone (see page 140). The brightest of the stars form the familiar constellations in our night skies.

The stars are so very far removed from Earth that their distance is difficult to measure. Distances up to about 70 light-years can be established with relative accuracy by using a method similar to triangulation, which is used by surveyors. Greater stellar distances are determined indirectly. Because a star (like any other luminous body) appears fainter the farther away it is, scientists try to calculate a star's magnitude theoretically by applying methods based on physics. They then compare the results with the actual magnitude observed. However, this method is less accurate, and the distances given by different astronomers vary considerably, diverging sometimes as much as 100 percent, especially for very remote stars.

One of the most fascinating areas of research in modern astronomy is the study of the life history of stars. Stars are formed from huge interstellar gas clouds (see pages 141–42), in whose interior gases and particles of matter gradually concentrate. A star is born when, at the core of this concentrated mass, energy is first produced through the conversion of hydrogen, the most prevalent element in space, into helium. The star then goes on to spend the main part of its life in a stable state, shining with a steady light.

But at some point all the hydrogen is used up and the star enters its next life stage. It tries to find alternate sources of energy by transforming the helium it has produced. But this is accomplished only with difficulty. The interior parts of the star contract while the outer ones expand. The star swells up in size and becomes a red giant. Eventually, when all sources of energy are exhausted, the star's life comes to an end. The star goes through a final, cataclysmic reaction and then either dies in a spectacular supernova explosion or simply ceases to emit light (see pages 149–50). The outer layers go hurling into space as gas nebulas, and what remains is either a star corpse whose glow gradually fades away—a white dwarf (see Sirius, pages 146–47)—or a neutron star. Conditions in such a neutron star are beyond anything imaginable: The mass of the entire Sun can be compressed into a ball 6 to 12 miles (10–20 km) in diameter, and one cubic centimeter of matter weighs 10 million to 1 billion tons! The life span of a star depends largely on its initial mass. Heavy stars with more than 5 times the mass of our Sun live for a relatively short time, about 100 to 200 million years, whereas stars like our Sun have a life span of 7 to 10 billion years.

Galaxies
The Milky Way, also called Our Galaxy, has provided the word used to refer to the islands of stars (galaxies) that exist in countless numbers (at least 3 billion) in the vastness of space. Each galaxy is a star system composed of stars, gases, and dust. Large galaxies, such as the Milky Way, contain several hundred billion stars. Smaller ones, such as the Small Magellanic Cloud, still have several hundred million stars.

The galaxies are removed from each other by unimaginable distances—at least hundreds of thousands or, more commonly, many million light-years. The vast regions of space that intervene between them are practically empty. Except for the Magellanic Clouds in the southern sky and the Andromeda Galaxy in the northern sky, galaxies can be seen only through telescopes.

The Andromeda Galaxy, about 2.2 million light-years away and in many ways similar to our own galaxy, the Milky Way, is one of the most familiar celestial objects. It appears to the naked eye as a blurred patch, but through a telescope one clearly can make out an elongated spindle of light. It and the Magellanic Clouds are among the Milky Way's closest neighbors. The galaxies thus far mentioned plus 30 dwarf galaxies all belong to a cluster of galaxies known as the Local Group, which extends across about 5 million light-years. The Milky Way and the Andromeda Galaxy are the largest, most dominant members of the group. On the basis of their shape, galaxies are classified as being elliptical, spiral, barred spiral, and irregular. The most beautiful are the spiral shaped ones, of which the Andromeda Galaxy and the Milky Way are examples. In this type of galaxy, stars and gas clouds move around a large center with a great deal of mass, forming spiral-like arms the way a pinwheel does. The total number of stars in the Andromeda Galaxy is estimated at 400 billion and its diameter at about 100,000 light-years. In the past, this galaxy was referred to as the Andromeda Nebula; this name no longer is in use because the term nebula is now reserved for masses of gas inside the Milky Way. The technical name for this celestial body is M 31, which stands for the fact that this is the 31st entry in the catalog of celestial objects compiled by Charles Messier (see photo, page 26).

Halley's Comet during its last appearance close to Earth in 1986.

The Hyades
The constellation Taurus includes not only the Pleiades (see page 145) but also a second open star cluster, the Hyades. This cluster is easily visible to the unaided eye as a V west of the star Aldebaran. The name Hyades comes from the Greek and means rainy stars. The Hyades rise at dusk in autumn (see star maps for October). In earlier times they were considered harbingers of the rainy season.

The Hyades are composed of about 350 stars and are situated 150 light-years away from Earth. This makes them the second closest open star cluster, and it is this relative closeness to us that lends special interest to them for astronomers. The Hyades have become a kind of milepost for measuring distances in the universe.

In the case of the Hyades, a method for determining distance can be employed that works especially well and yields exact values without depending on special assumptions. The measurement of 150 light-years thus obtained is then used to gauge the magnitude of some stars whose spectra display certain characteristic features. If these features are then found in other stars, the difference between the brightness of the stars within the Hyades and that of the stars under actual observation is used to calculate the distance of these other stars (all radiant objects appear fainter and fainter the farther away they are). Thus an error in the assumption of the Hyades's distance seriously would affect the distance calculations of galaxies (see page 135) millions of light-years away. New attempts keep being made to measure the distance of the Hyades, but thus far the estimate of 150 light-years has proven fairly accurate.

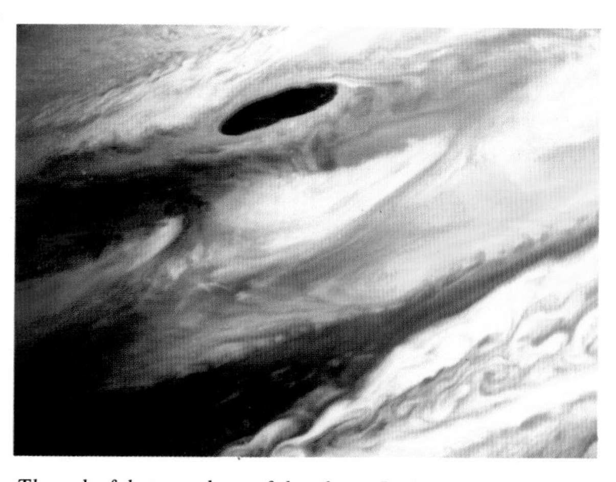

The colorful atmosphere of the planet Jupiter.

Jupiter

Jupiter is the largest planet of our solar system, larger and heavier than all the other planets taken together. Its name, the Roman version of Zeus who was the chief god and father of the Greek pantheon, does justice to Jupiter's position—the planet was dubbed the "royal star" by astronomers of antiquity.

The size of Jupiter is matched by the complexity of its system of satellites. Sixteen satellites have been identified thus far, four of which (discovered in 1614 by Galileo Galilei) are quite large, large enough to be seen through binoculars. Jupiter's makeup differs completely from that of the planets Mercury, Venus, Earth, and Mars, all of which are closer to the Sun. Jupiter is made up primarily of hydrogen gas, helium, and methane. It is thus a mass of gases, presumably surrounding a massive and dense core. It has a very extensive, thick atmosphere that is shaken by huge storms whose power and ferocity hardly can be imagined here on Earth. On the surface of Jupiter, one can detect huge swirls of clouds that can last for hundreds of years and grow as large as 25,000 miles (40,000 km) across.

Knowledge of the largest planet was added to considerably in 1979 when the American Voyager space probes passed by it. The instruments on the craft observed not only the chaotic weather patterns on Jupiter but also the planet's large moons. The moons turned out to be almost more interesting than Jupiter itself. This is true especially of Io, the closest of the satellites, on whose surface the first active volcanoes in the solar system were discovered (discounting the ones on Earth). We learned more about Jupiter when the spacecraft Galileo reached the giant planet in December of 1995 after a six-year journey. Galileo went into orbit around Jupiter and studied it and its moon for almost a year.

Basic Data of Jupiter

Distance to the Sun:	484 million miles (779 million km)
Period of revolution about the Sun:	11.9 years
Diameter:	88,673 miles (142,790 km)
Period of rotation:	9 hours, 55 minutes
Mass:	318 times that of Earth

M

The Magellanic Clouds

Out of a total of at least 3 billion galaxies in the universe, only three can be seen with the naked eye. They are the Andromeda Galaxy, M 31, in the northern sky and the two Magellanic Clouds in the southern sky. Whereas the Andromeda Galaxy (see page 135) appears as nothing more than a small, fuzzy blob in the autumn sky, the two Magellanic Clouds can be seen distinctly any night of the year from any point south of 20 degrees south latitude. They are near enough to the south celestial pole to be circumpolar in that area (see North Pole, South Pole, and Equator on page 9).

Both Magellanic Clouds, which look almost like pieces that had broken off the Milky Way, are about 160,000 light-years distant and are thus the closest galaxies to our own, the Milky Way. Together with the Andromeda Galaxy, the Milky Way, and a few other smaller galaxies, they form a kind of "galaxy cluster" that moves through space together, the so-called Local Group. Astronomers count the two Clouds among the irregular dwarf galaxies. But this has to be understood as an astronomer's term, for the Greater Magellanic Cloud contains about 6 billion times—and the Lesser one about 1.5 billion times—the mass of the Sun in the form of innumerable stars as well as gas and dust nebulas (see photos, pages 78 and 152–53). The two Magellanic Clouds form a stunning sight when viewed through binoculars. At observatories on the southern half of the globe they are *the* object of observation, and their study already has helped solve many astronomical problems. In January 1987, a supernova flared up in the Great Magellanic Cloud; it was the first supernova to be visible to the naked eye since the year 1604 (see pages 149–50).

Mars

Mars owes its name to its reddish color. It was named after the Roman god of war. The reddish color is due to Mars's extensive deserts covered with red sand.

In many respects, Mars resembles Earth. It has seasons like those of Earth, and it rotates around its axis in about the same amount of time as Earth does. It has a thin atmosphere, which is sometimes disturbed by quite violent weather with huge sand storms sometimes blocking the view of as much as half the planet's surface. Mars's polar caps, again like Earth's, are white, though most of the ice melts during the Mars summer. Because of these similarities to Earth it was thought in the last century that Mars might be inhabited, and to this day stories are being published in which Martians figure prominently.

Modern space exploration has, however, put an end to the belief that Mars might be inhabited. The only other planet that has been approached by space probes and vehicles as often as Mars is Venus. The culmination of the efforts to explore Mars came in 1976 when Viking I and Viking II landed on Mars and spent many months taking pictures of the planet's surface and picking up soil samples (see photo, page 139). The findings give no indication whatsoever of the existence of any form of life on Mars, no matter how primitive. The photographs show Mars to be covered with a great many craters and also with strange rills that look like rivers. It seems as if there had once been water on the surface of Mars many millions of years ago.

Mars, which is only half the size of Earth, can be observed only with difficulty. All that can be seen through a telescope at high magnification are light and dark areas. They were all that cartographers had to go on to construct maps of Mars before the age of space exploration.

Basic Data of Mars

Distance to the Sun:	142 million miles (227.9 million km)
Period of revolution about the Sun:	687 days
Diameter:	4,220 miles (6,794 km)
Period of rotation:	24 hours, 37 minutes
Mass:	.11 times that of Earth

Mercury

The winged messenger of the Greek gods lent his name to the planet closest to the Sun. The people of antiquity had already noticed that of all the planets Mercury moves across the sky with the greatest speed. It can be seen only for about a week at a time in the evening dusk or for about a week in the morning dawn, low in the horizon (see table, pages 114–15).

Mercury's speed is due entirely to its proximity to the Sun. Of the nine major planets of the solar system, Mercury is the one nearest to the Sun. Because the Sun's gravitational pull is especially strong there, Mercury has to move fast in order not to be

M

drawn off its course and plunge toward the Sun. The great closeness to the Sun also affects the planet's surface drastically. On the side turned toward the Sun the temperature can rise to 400° C, whereas on the side facing away it drops below minus 200° C. The temperature difference is so extreme because Mercury has no atmosphere like Earth's that could act as a buffer.

Thus far only one space probe has approached Mercury. This was Mariner 10, which passed very close to Mercury three times in the years 1974 and 1975, as close as 203 miles (327 km) at the last flyby on March 16, 1975. The cameras transmitted thousands of sharp television pictures to Earth, which show the planet's surface to be a stony desert pockmarked by craters. Mercury looks very similar to the Moon. On the basis of the photographs taken by Mariner 10, detailed maps of Mercury have been made on which many craters are named after astronomers and other famous personages. Some of Mercury's craters are extremely large; thus the crater Beethoven measures 388 miles (625 km) across. But there are also large, smooth plains on the planet, such as the Calores Basin, which extends across about 800 miles (1,300 km). Presumably the Mercury craters were caused, like those of the Moon, by the impact of large meteorites (see pages 138–39) many million or even billion years ago.

Basic Data of Mercury

Distance to the Sun:	36 million miles (57.9 million km)
Period of revolution about the Sun:	88 days
Diameter:	3,029 miles (4,878 km)
Period of rotation:	58 days
Mass:	.05 times that of Earth

Meteors and Meteorites

On a clear night, one frequently sees brief streaks of light in the sky—the shooting stars or meteors. Their sometimes striking trails that shoot across the sky at lightning speed are caused by tiny particles of dust, often the remains of comets, penetrating into Earth's atmosphere.

The air that is set ablaze as a meteor enters the atmosphere appears to us as the trail of a shooting star. Even a grain of dust less than one gram in weight can cause a meteor that looks brighter than the brightest stars.

Most shooting stars light up for only a fraction of a second. But there are also some, called fireballs, that are extremely bright and even produce sound. When the piece of matter that enters Earth's atmosphere from outer space is quite large, it may not be completely incinerated in its passage through the air but may survive and fall to Earth. These fragments from outer space are called meteorites. Unfortunately it is impossible to predict the arrival of meteorites or the appearance of meteors. Their occurrence is a matter of pure chance, although there are two interesting general patterns.

One is that meteors are always more numerous in the morning than in the evening, by

Meteor Showers

Name	Period of Visibility	Maximum	Number of Meteors per hour	Origin
Quarantids (Boötes)	1/1–1/4	1/3	40	unknown
Lyrids	4/12–4/24	4/22	15	Comet 1861 I
May Aquarids	4/28–5/21	5/5	20	Halley's Comet
July Aquarids	7/25–8/10	7/29	20	unknown
Perseids	7/20–8/19	8/11	50	Comet 1862 III
Orionids	10/11–10/30	10/20	25	Halley's Comet
Leonids	11/14–11/20	11/17	15 (variable in 1966 over 100,000!)	Comet 1866 I
Geminids	12/5–12/19	12/12	50	Asteroid Phaeton?

M

Olympus Mons, the largest volcano on Mars, photographed by Viking I.

roughly a factor of four. This is because in the morning we face in the direction of Earth's movement. Consequently our planet, racing along its orbit around the Sun, literally sweeps up more dust from space than it does in the evening when we look out in the direction opposite to Earth's movement, watching, so to speak, the wake of Earth's progress. The second predictable factor is that there are certain times of the year when more meteors than ordinary are produced—what are called meteor showers.

Meteor Showers

When comets disintegrate (see page 133) and dissipate their dust particles into space, these particles continue to orbit along the comet's path around the Sun. Every time Earth cuts across such a dense trail of dust, an unusually large number of particles enters Earth's atmosphere, causing many meteors to appear. These are the meteor showers or swarms.

Just as it looks to us, when we drive along a highway, as though the trees along the side of the road were flying toward us from a distant point ahead and whizzing past the car's side, so the meteors of a swarm seem to emanate from one spot in the sky and scatter from there in all directions. The name of the constellation from which a given meteor shower seems to originate is used in that shower's designation.

At the present time, the Perseids are the most prolific meteor showers. Because they are most plentiful in August, on the day of St. Lawrence, they are also called the Tears of Saint Lawrence. The Quadrantids, which appear in early January in the area of Boötes, are named for Quadrans (the Quadrant), an ancient constellation that is no longer recognized.

The most interesting meteor showers are the Leonids. They appear annually, but once every 33 years Earth traverses an especially heavy concentration of dust particles, and then meteors show up in such amazing numbers that the whole sky appears lit up. This happened on November 12, 1799 (observed by Alexander von Humboldt in South America), on November 12, 1833, and again on November 14, 1866, when 7,200 shooting stars per hour were observed. But in 1899 the Leonids failed to appear, and they were not seen again until November 14, 1966, when the most massive shower of meteors of all time was witnessed in the American Midwest, with at times as many as 40 meteors per second (a rate that translates into 144,000 per hour!). How the Leonids will behave in the future is hard to predict, but some observations from islands in the Atlantic showed hourly rates of thousands in November of 1998.

M

The center of the Milky Way—an aggregation of innumerable stars.

The Milky Way
One of the most striking celestial objects is the band of light, irregular in outline and broken up in places by dark areas, that arches across the sky. The Milky Way's band of muted luminosity fascinated the ancient astronomers and has given rise to the most diverse mythological explanations (see pages 40–41 and 48–49). The best known of these old legends still provides the basis for the modern technical name for the Milky Way, namely, the Galaxy (*gala* is the Greek word for milk).

The Galaxy, or Milky Way, is the cosmic home of our Sun and Earth in the universe. It is a huge system of stars in the shape of a disk. Regarded from a point outside of it, the Milky Way would look much like the Andromeda Galaxy, which is a sister galaxy of the Milky Way (see page 135). It has a diameter of 100,000 light-years. Millions of stars revolve in huge orbits around an immense central section that must contain a mass at least 70 million larger than our Sun. For reasons not yet understood, the stars are arranged in spiral arms. Looked at from above, the Galaxy therefore would look like a pinwheel of astronomical proportions. Our solar system is located in the outer reaches of the Galaxy, about 28,000 light-years from the center. At this distance, our solar system takes about 200 million years to circle once around the Galaxy's center. The spiral arms of the Milky Way Galaxy move around this center similarly to the way Earth and the other planets revolve around the Sun.

The stars are largely concentrated along the plane of the Milky Way's disk, and because Earth also lies on this plane, our Galaxy's stars seem to be arranged in a band around Earth. We are able to see only the closest ones individually; the millions of stars in the Galaxy's spiral arms blur together to form a luminous area. Because there are also many gas and dust clouds located in the spiral arms, this band appears jagged and in places seems to have holes and fissures. Some examples of this are the Coalsack in the constellation of the Southern Cross (see pages 84–85) and the area of the constellation Cygnus, where the Milky Way seems to divide in two separate strands (see pages 44–45).

The band of the Milky Way is so faint that excellent viewing conditions are required for its observation. Preferably the night should not be moonlit.

Mira
The Wonderful Star (*mira* is the Latin word for wonderful) is part of the constellation Cetus and was the first variable star to be discovered (see pages 74–75). It was first observed as early as 1596 by the East Frisian pastor David Fabricius, who assumed he had discovered a new star but then could not find it again until 1606 and therefore named it Mira. It was not realized until 1638 that this was a variable star

that at the peak of its luminosity is easily visible to the naked eye but can be discerned only with the aid of a telescope in its faintest phase. Mira, which is 220 light-years away from Earth, belongs to the class of red giants. Its brightness fluctuates regularly in periods of about 330 days or approximately 11 months. Mira is a pulsating star; that is, it alternately expands and contracts, thus changing the size of its light-radiating surface. At maximum brightness, Mira is about 500 times the size of our Sun; at minimum brightness, about 350 times. Today a total of about 5,000 stars are known that fluctuate in magnitude for the same reasons and therefore are called Mira-type variables.

The Moon

Most of the planets in our solar system are accompanied by satellites or moons. The word moon used in this sense obviously derives from our Moon, Earth's traveling companion.

Earth has only one satellite, but it occupies a special place in the solar system because it is one of the largest satellites when regarded in terms of the size of the bodies around which they orbit. The Moon's mass is about 1/81 that of Earth, whereas the satellites of the other planets (except for Pluto) have less than 1/1000 of the mass of the planet they circle. For this reason it is easier to think of Earth and the Moon as a kind of double planet traveling around the Sun together.

The Moon is the closest heavenly object to Earth and the only one, apart from the Sun, that appears as a disk to the naked-eye observer. One clearly can make out dark and light spots on its surface that add up to something that can be interpreted to look like a face (the "man in the moon"). The dark spots are low-lying areas on the Moon's surface that are covered with dark lava. These are called the Moon's seas or maria and have very few craters. The light areas indicate areas of greater altitude that are covered with granite-like rocks and dotted with a multitude of craters. They are sometimes called the Moon's continents even though there is no water on the Moon. These names go back to the time of the first examinations of the Moon through telescopes.

Even a look through binoculars reveals how densely some of the Moon's surface is covered with craters. One also can see mountain ranges, such as the Lunar Apennines and the Lunar Alps. The Moon's craters, some of which can measure as much as 125 miles (200 km) across, are named after great scientists and other famous persons in world history. The names of the Moon's seas are more imaginative. The Moon's appearance varies as its position in relation to Earth and the Sun changes. These moon phases (for details on how they are produced, see pages 105–07) also affect significantly what is seen when observing the Moon through binoculars. At full moon, the moon craters and mountains do not cast any shadows, so that they do not stand out three-dimensionally. The craters, mountains, and maria are easiest to see in the border area between the light and dark sides of the Moon. Look for them in the evening and preferably during the first quarter phase (see Table of Lunar Phases, pages 106–07).

The same side of the Moon always is visible because the Moon takes the same amount of time to rotate around its axis and to revolve around Earth. But now, thanks to photographs taken by many spacecraft circling the Moon, it is known exactly what its far side looks like. To date, 12 astronauts aboard six spacecraft, the American Apollo missions 11 to 17, have landed on the Moon, as have about two dozen unmanned space vehicles. The Moon therefore is the most thoroughly explored celestial object.

Basic Data of the Moon

Average distance from Earth:	239,000 miles (384,400 km)
Period of revolution around Earth:	27⅓ days (sidereal month; that is, one orbit relative to a fixed point, such as a distant star) 29½ days (synodic month; that is, one cycle from new moon to new moon as seen from Earth)
Diameter:	2,160 miles (3,476 km) (.27 times Earth's diameter)
Mass:	1/81 of Earth's mass

Nebulas

The space between the stars is not a void. It is filled with so-called interstellar material; however, the density of this material is so extremely low that the emptiest vacuum that can be created on Earth would seem like a packed hall by comparison. In one liter capacity of interstellar space there are at most 50,000 to

100,000 atoms, whereas the same unit of capacity filled with Earth's air contains 30 sextillion atoms (that is a 3 followed by 22 zeroes!).

Ninety-nine percent of the interstellar material is made up of gases—hydrogen and helium atoms—the other one percent consists of dust particles. Together they form the cosmic nebulas, which are some of the loveliest and most colorful phenomena in space. When there are bright stars in the vicinity of a gaseous nebula, the energy-rich ultraviolet light from the stars causes the gases to emit light of their own. The gases then glow in various colors and sometimes chaotic shapes, whereas the dust clouds project the most fantastic silhouettes against these brilliant walls of light. The dust clouds intercept and absorb the light rays that are on their way toward Earth.

Viewed without optical magnification, the cosmic nebulas appear only as small, fuzzy patches, but their striking shapes can be glimpsed even through binoculars and are, of course, much more impressive when seen through a telescope. The colors show up only on photographs taken with huge telescopes and long exposure times. The three most famous nebulas, which are marked in the star charts, are:

- **The Great Nebula in Orion or M 42** This is the brightest nebula in the sky and is considered by many astronomers to be the greatest celestial showpiece. With a diameter of 100 light-years it covers a sizable piece of the sky. To many viewers who glimpse it through a telescope, it seems like chaos in visible form. M 42 is 1,500 light-years distant from us and owes its great luminosity to four extremely bright stars at its center. These stars are known as the Trapezium in Orion because they seem to form the corners of this geometrical figure (see photo, page 52).
- **The Lagoon Nebula or M 8** The English astronomer John Flamsteed discovered this galactic nebula in the constellation Sagittarius in the year 1747. It shines forth among the stars of the Milky Way with an intensive red, the color emitted by hydrogen gas at a temperature of about 10,000 degrees. A dark cloud cuts across this brightly shining mass of gas and creates the impression of a lagoon (see photo, page 5). There is enough gas in the Lagoon Nebula to form 200 suns. The nebula is about 3,000 light-years away.
- **Eta Carinae Nebula** This is one of the most beautiful objects in the southern region of the Milky Way. A huge hydrogen cloud surrounds the star Eta Carinae at a distance from Earth of about 6,000 light-years. It is to this star of superlatives, which was for a long time taken to be the brightest and heaviest star of the entire Milky Way Galaxy, that the nebula owes its luminosity. This is because Eta Carinae radiates at least 4 million times more light than the Sun and thus excites the gas atoms within a radius of several light-years to light up.

Today Eta Carinae, the star that gave rise to the nebula, is no longer visible to the naked eye. But during the last century, it was among the brightest stars in the sky. In 1843, it was the second brightest, right after Sirius, but after that its magnitude declined rapidly until, by 1865, it had become an object visible only through a telescope, and it has remained more or less unchanged since then. The reason for this strange change is still a mystery today. The most recent measurements seem to indicate that Eta Carinae is actually a multiple star consisting of four or five stars, and many astronomers see in it a possible candidate for the next supernova explosion in the Milky Way sometime within the next hundred years (see pages 149–50).

Northern and Southern Lights
In the high latitudes both north and south, around the magnetic poles, one often can observe amazing light shows in the sky at night. Ghostlike greenish and reddish beams of light and entire walls of light chase across the sky or may hover overhead for hours without moving. These auroras, or northern and southern lights, are caused by the Sun, which emits a steady stream of charged particles into space. Earth's magnetic field directs these particles toward the polar regions, where they collide with molecules in the air and excite them to glow. Northern (and southern) lights are especially frequent at the peak of sunspot cycles.

Novas
From time to time stars blaze up unexpectedly in the sky and become so bright that they are visible to the unaided eye for a few days or weeks before sinking into obscurity again. They undergo such a sudden and dramatic increase in magnitude that they might be taken for new stars (*nova* is Latin for new). In fact they exist both before and after their spectacular performance. Novas are variable stars (see pages 150–51) whose fluctuations in magnitude do not follow a more or less regular pattern but happen abruptly and are some-

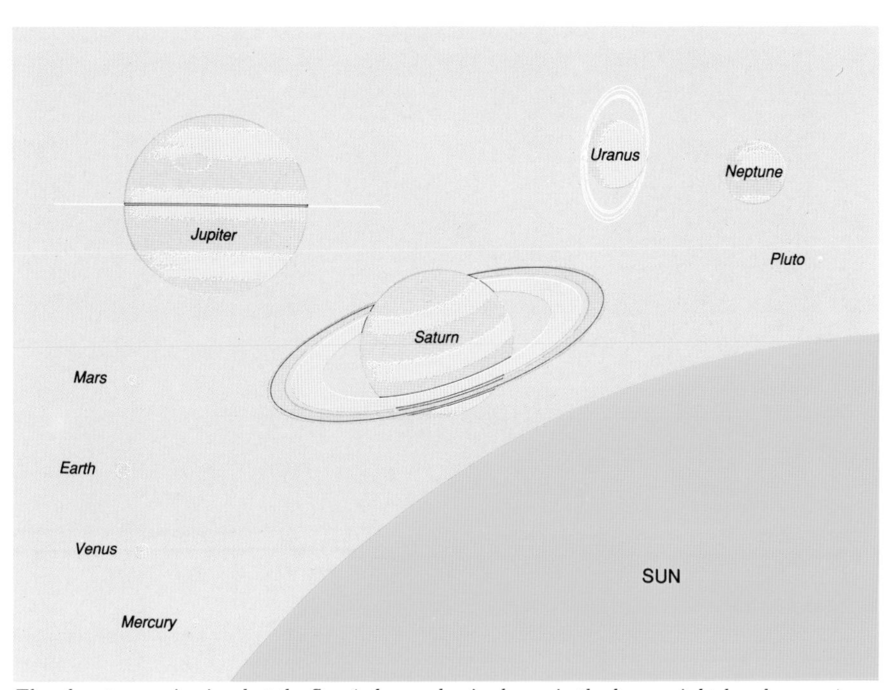

The planets vary in size, but the Sun (whose edge is shown in the lower right-hand corner) dwarfs them all.

times of stupendous dimensions. For example, in the year 1975 the nova in the constellation Cygnus increased in magnitude by a factor of 40 million within a few days (see pages 44–45). Then, over a period of several weeks, it gradually returned to its previous level. There are also stars in which nova explosions have been observed more than once.

Astronomers suspect today that all novas are binaries (see page 131). One of the two stars is a very small, hot, and dense white dwarf (see pages 146–47); the other, a large, cool, reddish star. The two are so close to each other that matter keeps being drawn from the larger to the smaller one. Eventually so much matter from the larger companion accumulates on the surface of the white dwarf that it begins to ignite spontaneously. Hydrogen fuses into helium by means of the same process that goes on in the interior of stars, including our Sun (see page 149). But here the nuclear fusion is not controlled as it is in the Sun but instead takes the form of an explosion—similar to the explosion of hydrogen bombs here on Earth, which function on the same principle of fusing hydrogen into helium. During the gigantic explosion of a nova, the hot hydrogen layer enveloping the small star is shed, and the whole game can start over until the larger star has lost all its outer matter to the smaller one.

Novas are to be distinguished from supernovas (see pages 149–50).

The Planets

The planets of our solar system appear in our sky as nothing more than dots of light just like the fixed stars (see pages 133–34). Yet the difference between these two groups of heavenly objects could hardly be greater. Planets are cold celestial bodies that revolve around the Sun and are illuminated by it. They are much closer to Earth than any stars, their distance ranging from 24 million miles or 38.3 million km (Venus at its closest) to 4.6 billion miles or 7.5 billion km (Pluto at its farthest). The closest star to us, Alpha Centauri (see pages 130–31), is about 27 trillion miles (43 trillion km) away.

Astronomers suspect today that many stars have planetary systems. However, despite many

P

The planet Saturn in a photograph taken by Voyager I with colors added by computer.

tries, no one has been able thus far to supply conclusive proof that this is so. We therefore know only one planetary system at this point, namely that of the Sun. It consists of nine major planets and a large, indeterminate number of minor planets or asteroids. These minor planets can be seen only through a telescope and, because of their small size, show up even there only as tiny dots, like stars. Most of them orbit around the Sun between Mars and Jupiter. Today the paths of about 4,000 asteroids, almost all of which have been given names, are known. In earlier times they were named after legendary figures, now mostly after astronomers and other natural scientists.

By contrast, clear details can be observed in the surface of the major planets when they are examined through a telescope. Still, exact knowledge of the planets had to wait until the age of space exploration, when space probes photographed and measured the planets from up close and transmitted the data back to Earth. At this point in time, all the major planets have been visited by space probes with the exception of Pluto, which is farthest away from the Sun. The most sensational accomplishments were scored by Voyager II, which started out in 1977 and traveled until 1989, making an almost incredible journey of 3.7 billion miles (6 billion km) through the solar system, in the course of which it reached Jupiter, Saturn, Uranus, and Neptune.

Three of the major planets cannot be seen with the naked eye but are nevertheless also very interesting. They are:

- **Uranus** It was discovered in

Basic Data of the Planets	**Uranus**	**Neptune**	**Pluto**
Distance from the Sun:	1.7 billion miles (2.8 billion km)	2.8 billion miles (4.5 billion km)	3.7 billion miles (5.9 billion km)
Period of revolution about the Sun:	84.6 years	165.5 years	248 years
Diameter:	31,500 miles (50,800 km)	30,200 miles (48,600 km)	1,860 miles (3,000 km)
Mass:	14 times that of Earth	17 times that of Earth	.002 times that of Earth

1781 by the German-English astronomer William Herschel. Uranus greatly resembles Jupiter in physical structure, consisting primarily of gases, but its mass is considerably less. In pictures taken by Voyager II on a flyby in 1986, Uranus revealed itself as a quiet planet with surprisingly few notable features in its dense atmosphere, which shimmers in a bluish color.

• **Neptune** The discovery of Neptune in 1846 was one of the greatest triumphs of natural science in the nineteenth century. The English astronomer J. C. Adams and the French astronomer U. C. Leverrier independently from each other had predicted the existence of the planet. Both had observed that Uranus did not orbit around the Sun along the path it should have followed according to astronomical calculations. They consequently theorized that the gravitational force of another planet must be acting on Uranus, causing it to deviate from its path, and, basing their calculations on the observed deviations, they predicted the location of the missing planet. Neptune was found exactly where the scientists had said it would be.

Very little was known about this planet before August 1989 when Voyager II reached it. To everyone's surprise, the space probe's photographs showed Neptune to be a very active planet—far more active than Uranus—with a great deal of turbulence and large bands of clouds in its atmosphere. Neptune, like Uranus, resembles Jupiter in consisting primarily of gases.

• **Pluto** The American Clyde Tombaugh spotted Pluto in 1930. But it was not until 1978 that a moon, Charon, was discovered orbiting around Pluto. From this satellite's path astronomers were able to calculate the mass and estimate the diameter of Pluto (see Binaries, page 131). Pluto turned out to be by far the smallest and lightest planet of the solar system. Many astronomers suspect that Pluto might once have been a moon of Neptune.

The Pleiades

The seven daughters of the giant Atlas (see pages 54–55) not only turned Orion's head with their beauty but also were transformed by the gods into one of the most beautiful objects in the sky. On a clear winter night, they are visible in the constellation Taurus as a faintly shimmering patch that looks very impressive when viewed through binoculars. The unaided eye can make out from six to thirteen stars, seven of which bear the names of the seven sisters: Alcyone (the brightest of the stars), Maia, Asterope, Tageta, Celaeno, Electra, and Merope. The Pleiades are the most well-known open star cluster (see pages 148–49). This cluster is made up of about 130 stars spread out over an area of 30 light-years. The Pleiades are 410 light-years away from Earth and are one of the younger star clusters. They are thought to be about 60 million years old.

The stars of the Pleiades are enveloped by wisps of tenuous nebulas that can be seen clearly in photographs taken with long exposure times. These nebulas are dust clouds that reflect the light of the stars constituting the Pleiades (see photo, page 147).

The Praesepe

The Praesepe, also called the Beehive, is another open star cluster in addition to the Pleiades and the Hyades (see page 135) that has a descriptive name (*praesepe* is Latin for manger). This star cluster is located in the constellation Cancer. Its technical name, which is the one used in the star maps, is M 44. Viewed through binoculars, the Praesepe is the second most impressive open star cluster after the Pleiades.

A total of about 500 stars are contained in the Praesepe, which is 520 light-years away from Earth. They are distributed over a space of 15 light-years and are approximately 400 million years old (see photo, page 25).

Procyon

Stars appear especially bright in our sky for one of two reasons: They are either exceptionally large and produce immense quantities of energy, as in the case of Betelgeuse (see page 131) and Rigel (see page 146), or they are very close to Earth, like Alpha Centauri (see pages 130–31) and Sirius (see pages 146–47).

Procyon in the constellation Canis Minor belongs to the second group. Without Procyon, this small constellation would hardly be worth mentioning. The name Procyon comes from the Greek and means before the dog. The reason for the name is that Procyon appears in the east shortly before its more striking companion, Sirius in Canis Major. Procyon is a star similar to our Sun, but about six times brighter and twice its size. Astronomers were able to determine its mass more precisely than that of Sirius because Procyon is a binary (see page 131). Its partner revolves around it in 40 years and is a white dwarf. In this respect, too, Procyon resembles its big brother, Sirius. The mass of Procyon has been estimated at 1.7 times that of the Sun; that of its much smaller companion star, at .65 the mass of our Sun.

R–S

Rigel

Rijel Jauzah al Yusra (the left leg of the giant) is what Arabian astronomers called the bright white star in Orion (see pages 54–55). This name survives today in the much abbreviated form Rigel. Rigel is a star of truly astronomical dimensions. Of the ten stars that appear brightest in the sky, it is the one farthest removed from Earth; astronomers estimate the distance variously from 900 to 1,300 light-years. The fact that Rigel is nevertheless so prominent in our sky means that it must generate immense amounts of energy. Astronomers class it as a supergiant, with a surface temperature of 12,000 degrees and an absolute magnitude equal to that of 57,000 suns. Other basic data about it includes a mass 30 times that of the Sun and a diameter 19 times greater. It shines 40 times brighter than Venus and is almost a fifth as bright as the Moon at full moon.

Saturn

The most beautiful of the Sun's planets is no doubt Saturn. Although it is known now, thanks to the flybys of the great space probes, that Jupiter, Uranus, and Neptune also have rings, Saturn's are by far the biggest. Even through a small telescope—though not binoculars—Saturn's highly impressive ring system is visible at a magnification of 40 or above.

Saturn is encircled by several rings. These are not solid but are made up of huge numbers of dust particles and small, gravel-like rocks that orbit around the planet. It is the great distance from Earth that fuses these countless pieces of material into an image of solid rings. The pictures taken by Voyager I and Voyager II, which passed by Saturn in November 1980 and August 1981, show the exact composition of the planet's ring system. The rings are thought to be the remains of a Saturnian moon that approached the mother planet too closely millions of years ago and was torn asunder by it. Saturn has other moons, some of which were not known until the space probes revealed them. One of these satellites, appropriately named Titan, vies with Jupiter's moon Ganymede for the rank of the largest planetary companion in the solar system. It, too, can be seen from Earth—with the aid of binoculars—and has a diameter of 3,240 miles (5,220 km).

Like Jupiter, Saturn consists of very light gases, primarily hydrogen, methane, and ammonia. The density of its material is so light that the planet would float on water. By contrast, Earth, which, like Mars and Venus, consists of metals and heavy minerals, would immediately sink to the bottom in such an imaginary experiment. Again as in the case of Jupiter, the outer surface of Saturn's atmosphere moves and shows bands of clouds and huge disturbances. But these atmospheric activities clearly are minor when compared to the tremendous storms that take place on Jupiter.

Astronomers hope to obtain more detailed information about Saturn primarily from the flight of Cassini, a space probe launched toward Saturn October 15, 1997. This spacecraft is scheduled to reach the vicinity of the distant planet in the summer of 2004 and then veer toward it. It will release a small module with scientific instruments that will parachute toward and land on Saturn's moon Titan.

Basic Data of Saturn

Distance to the Sun:	890 million miles (1,432 million km)
Period of revolution about the Sun:	29.6 years
Diameter:	74,500 miles (120,000 km)
Period of rotation:	10 hours, 40 minutes
Mass:	95 times that of Earth

Sirius

Sirius, the most famous of the stars, owes its great apparent magnitude primarily to the simple fact that it is only 8.7 light-years away from Earth. It sends out 23 times as much light as the Sun and is 1.8 times larger than it. Compared to stars like Rigel or Betelgeuse (see page 131), Sirius is relatively small. But the history of the brightest star is unusual and remarkable. In ancient Egypt, Sirius enjoyed godlike veneration because its appearance in the sky heralded the flooding of the Nile (see pages 56–57), and in ancient Rome, dogs were sacrificed to it. Astronomers in those early times consistently described Sirius as being reddish, indeed, redder than the planet Mars. Today its light is pure white, as a glance into the sky will show.

How can a star change its color over a period of barely 1,500 years? This question has yet to find a convincing answer. It is one of the great riddles of stellar astronomy, the study of fixed stars. For although stars do pass through different stages of development, such unmistakable, clearly visible

S

The star cluster in the Pleiades, the most beautiful in the sky.

changes as a shift of color from red to white require, according to all accepted theories, several hundred thousand years to complete, not a mere one and a half millenia.

Perhaps Sirius's mysterious change of color has something to do with its companion, for Sirius is a binary. As early as 1844, the German astronomer Friedrich Bessel had noted that Sirius does not move across the sky in a straight line (fixed stars do move, see page 134) but follows a serpentine path instead. Bessel concluded from this that Sirius must have an invisible companion whose gravity affects the behavior of its large partner. It took until 1862 before this companion, now officially named Sirius B, actually was discovered through a telescope as a very faint dot of light next to brightly shining Sirius A.

The actual discovery of Sirius B, or the Pup, as it is also known, presented a puzzle to astronomers. For on the basis of the two stars' movements (see Binaries, page 131–132) they had calculated that Sirius A must be 2.36 times as heavy and Sirius B .98 as heavy as the Sun. Because, however, Sirius B appeared so much fainter than its big brother (although its surface is extremely hot), it had to be much smaller, only about 18,000 miles (30,000 km) or about twice Earth's diameter. Such a large amount of matter concentrated in so little space meant a density that was beyond what anyone could imagine. One cubic centimeter of the matter making up Sirius B would weigh over 300 pounds (150 kg)! Thus Sirius B became the first example of a new type of star to be discovered, the white dwarfs. The characteristics of white dwarfs are extremely small size (the smallest one known thus far is only half as big as the Moon), high surface temperature, and incredible concentration of the matter composing them. Astronomers say that the matter in a white dwarf has degenerated. White dwarfs are dying stars that no longer generate light and therefore gradually fade. They are the remains of stars whose "hearts," that is, the generation of energy inside them, have stopped. One might say that they are star corpses.

Star Clusters

Stars do not exist in the universe in random isolation. Instead they are organized in systems or galaxies within which, under the force of gravity, they revolve around each other or move along common paths. Thus all the visible stars are part of the Milky Way Galaxy (see page 140), and many of them have companions and are thus double or multiple

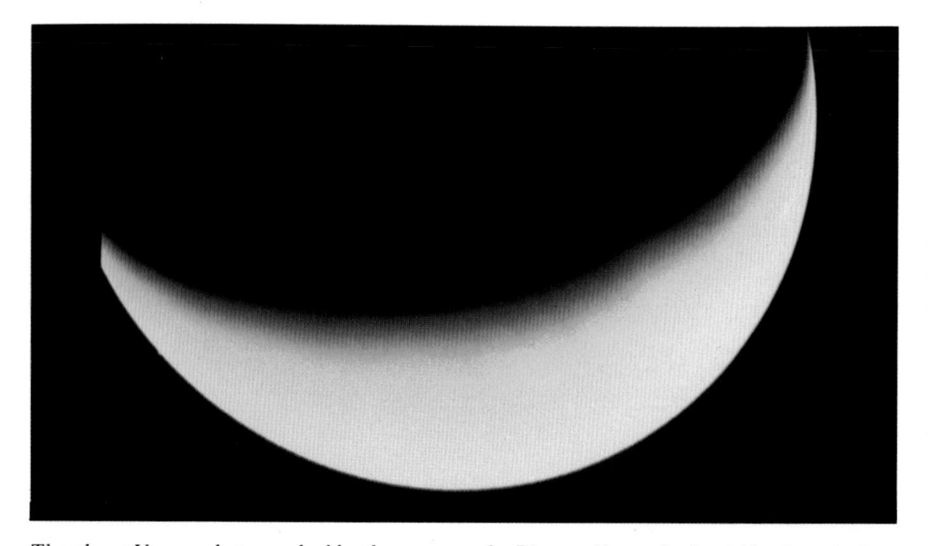

The planet Venus, photographed by the space probe Pioneer-Venus, looks sickle-shaped when on the same side of the Sun as Earth.

star systems (see page 131–132), or they form aggregations that include several dozen to many hundred stars, the star clusters. Our Sun, which is neither a member of a star cluster nor part of a multiple star system, is something of an exception in the Milky Way.

To date, astronomers have discovered about 1,000 star clusters in the Milky Way Galaxy alone. Depending on their appearance they are classified as either open star clusters, in which the stars are still individually recognizable in spite of their large number, or as globular star clusters, in which the stars are packed so close together that they appear as single white patches.

The three most famous star clusters are the Pleiades and the Hyades in the constellation Taurus (see pages 145 and 135) and the Praesepe, or the Beehive, in Cancer (see page 145). Apart from these, the following can be seen with the naked eye:

- **H and χ (chi) Persei** The constellation Perseus boasts not just one but two star clusters that are right next to each other and can be seen together through binoculars. Because they look almost like two stars to the naked eye, they have typical star names (see the Constellations and the Zodiac, starting on page 8). H Persei is somewhat more spectacular, having brighter stars. Both clusters are 8,000 light-years away from us, and they each contain 300 to 350 stars.
- **IC 2602 in Carina** The stars of this cluster also are referred to as the southern Pleiades. Their prosaic technical name indicates that they are the 2602nd entry in the Index Catalogue, a compilation of star clusters and nebulas. At a distance of 700 light-years, they are very close to Earth in astronomical terms. This cluster includes one especially bright star, Theta Carinae, with whose help it is easy to spot the cluster.
- **M 7 in Scorpius** is also an open star cluster. It is situated in a particularly starry region of the sky and therefore appears to the naked-eye observer as nothing more than a small, fuzzy patch in front of the luminous band of the Milky Way. M 7 is an aggregation of about 60 stars about 780 light-years away. This cluster was studied closely by the astronomer Charles Messier in the spring of 1746 and included as the seventh and the most southerly object in his catalog.
- **Omega Centauri** is the brightest of all the globular star clusters. The fact that in spite of a distance of 15,000 light-years this cluster can be recognized quite clearly as a star-like, fuzzy spot is an indication of its great inherent luminosity. The total number of stars composing the cluster is estimated at 100,000. Globular star clusters like Omega Centauri play an important role in astronomy when it comes to determining the age of stars. Because all the stars in a cluster

must have originated at the same time, they give an exact picture of the stage of development different stars are at after the passage of a given span of time. In this way the age of Omega Centauri has been established at approximately 8 to 10 billion years (see photo, page 130).

The Sun

The Sun is the most important celestial body, without which there could be no life on Earth. Without the continual flow of solar energy, Earth would be a dead, ice-cold ball of rock with an average temperature of minus 270°C. But the Sun has been radiating immense amounts of energy toward Earth for several billion years in the form of visible light. The energy the Sun emits is enough to make all the water on Earth's surface—all the oceans, icebergs, rivers, and lakes—evaporate within 10 seconds! But because of the huge distance that separates the Sun from Earth, only about one half of one billionth of the Sun's radiation reaches Earth. This still amounts to 1.5 quintillion kilowatt-hours annually.

The Sun is the star that is closest to Earth. Just like all other stars, which are much farther away, it generates energy and produces its own light. Compared to other stars, it is a rather insignificant, small specimen with only average mass, luminosity, and energy production. All the other bright stars that can be seen well with the unaided eye are clearly larger and more luminous—in many cases dramatically so (see Antares, page 69, Arcturus, page 36, and Deneb, pages 44–45).

Still, in comparison to Earth, the Sun's dimensions are overwhelmingly large. It is assumed that the Sun derives its energy from a process of thermonuclear fusion that goes on at its core. There temperatures of around 16 million degrees and a pressure of about 200 billion atmospheres prevail. Under these conditions, hydrogen atoms (which make up most of the Sun's matter) fuse into helium, a process during which minute quantities of mass are converted into energy in accordance with Albert Einstein's famous equation: energy equals mass times the speed of light squared, or $E = mc^2$. The Sun thus "uses up" about 4 million tons of mass per second, but it is so staggeringly large that from the time of its infancy it has converted only about .03 percent of its mass into energy.

The Sun is the only star on whose surface we are able to detect specific features. The most famous of these is the sunspots, relatively cool regions in the Sun's surface that appear dark because the temperature there is lower. In these sunspots huge eruptions often take place, during which radiation and matter are flung out into space. These eruptions are responsible for the famous northern and southern lights on Earth. They also can lead to radio blackouts, and they affect the upper atmosphere in various ways, many of which are not yet completely understood.

The number of sunspots—and consequently the Sun's activity—varies. Sunspots come in cycles of about 11 years. The causes of these sunspot cycles and of many other processes taking place on the Sun's surface are still mysterious. The Sun is moving into a relatively inactive phase as the year 2001 begins. The last period of maximum sunspots was during 1999 and 2000; the next minimum is expected around the year 2005. A new maximum of activity accompanying high numbers of sunspots is due around the year 2011.

Basic Data of the Sun

Diameter:	870,000 miles (1.4 million km) (109 times that of Earth)
Distance from Earth (mean):	93 million miles (149.6 million km)
Mass:	2 septillion (2×10^{27}) tons or 333,000 times Earth's mass
Age:	ca. 5 billion years
Surface temperature:	5,800°C
Total energy output:	380,000 quintillion (10^{18}) kilowatts

Because of its tremendous importance for Earth, the Sun always has been an object of religious veneration. Many old cultures had sun gods, and even today the Sun is often invoked as a symbol of life and strength, as at summer and winter solstice celebrations in mid-June and mid-December, respectively. (For more on these dates and the seasons in general, see pages 13 and 104.)

Supernovas

Among the most spectacular but unfortunately also the rarest phenomena to be observed in the sky are the supernovas. A star suddenly blazes up, becomes incredibly bright and outshines all other stars for some weeks or months. In the previous millennium, only six supernovas visible to the naked eye occurred, namely in the years 1006 in the constellation Lupus (see pages 96–97), 1054 in Taurus, 1181 and 1572 in Cassiopeia (see pages 28–29), 1604 in Ophiuchus (see pages 66–67), and 1987 in the Large Magellanic Cloud (see page 137). Through telescopes, about another 400 supernovas have been seen in distant galaxies (see page 135).

Supernovas seem to come out of nowhere and were therefore dubbed "new stars" (from Latin *nova*, see page 142), but in effect these stars existed all along in a much more inconspicuous and much less bright form before blazing up so dramatically. Within a few weeks, such a star suddenly generates the equivalent of energy it would take our Sun billions of years to manufacture; its outer parts literally are torn apart and pour into space at tremendous speeds. Astronomers therefore think that the appearance of a supernova marks the death of an especially heavy star. When a star has exhausted all means of generating energy, it eventually collapses. In this sudden and chaotic collapse of the star's core, unimaginable quantities of energy are released. The innermost parts remain as a neutron star, whereas the outer ones survive as gas nebulas, thus enriching the interstellar material from which new stars may form later. But not all stars end life this way. Smaller ones like our Sun lose their light in a much less spectacular fashion.

There are still many unsolved questions concerning the phenomenon of supernovas. To some of these questions, the analysis of detailed observations made during the supernova in the Large Magellanic Cloud in 1987 is beginning to yield answers. One aspect of all this is particularly intriguing. The immense amount of energy involved in a supernova eruption releases various elements from the core of the dying star, elements that were formed there millions of years ago. In the course of the explosion new elements are formed as well, such as carbon, oxygen, and nitrogen, and metals, such as iron, nickel, and cobalt. All elements except for hydrogen, which still makes up 90 percent of all the matter in space today, were hurled into space a very long time ago by supernova explosions, and they are what is the foundation of life. We thus owe our existence to the supernovas of past astronomical ages.

Variable Stars

Many stars do not always exhibit the same brightness. Instead they change their magnitude in cycles anywhere from several hours to several years in length, but most commonly with a so-called period of up to 100 days. In many of these variable stars the magnitudinal fluctuations occur with considerable regularity, whereas in others they follow no clearly recognizable pattern.

There are two reasons for the fluctuations in brightness. Either the physical characteristics of the star change—it may be growing larger or smaller or its surface temperature may rise and drop—or the star is concealed by another star that stands in our direct line of vision—as the Sun is hidden by the Moon in a solar eclipe—and blocks the light of the star behind it from our view for a certain period. The stars that change in brightness for the first reason are called intrinsic variables, and the most famous one of them is Mira in the constellation Cetus (see pages 74 and 140). The second group consists of stars called eclipsing binaries. Their most famous representative is Algol, the Demon Star in the constellation Perseus (see pages 30–31). Eclipsing binaries also could be considered as a subgroup of the spectroscopic binaries (see page 132) whose component stars cannot be seen separately through a telescope. However, their existence is revealed not only by characteristic changes in their color spectrum but also by the orientation of their orbits, on which they can mutually obscure each other for the viewer on Earth. Only few variable stars can be detected by the naked-eye stargazer, which is why, apart from Mira and Algol, only the following are marked in the star maps:

• **Delta Cephei** Although this star was known to the astronomers of antiquity because of its brightness, the changes in magnitude were not noticed until 1784, when the Englishman Goodricke discovered the star's variability. This is all the more surprising because the fluctuation can be detected with the unaided eye, as can that of Algol. Delta Cephei is a huge star, one of the so-called supergiants. It shines 3,300 times more brightly than the Sun. The fluctuation occurs in an extremely regular cycle of 5 days, 8 hours, and 48 minutes as Delta Cephei alternately expands and contracts, with a resulting change in diameter of 6 percent. The surface temperature and the radiating surface area fluctuate as well, and consequently, of course, the star's brightness. Delta Cephei is an intrinsic variable. It is about 1,200 light-years away from Earth.

• **L 2 Puppis** is also an intrinsic variable. It resembles Mira (see pages 140–41). Its brightness fluctuates within a period of 141 days. At a distance of 200 light-years it shines with a mean intrinsic luminosity that is 200 times greater than the Sun's.

• **Epsilon Aurigae** is an eclipsing binary and one of the most mysterious star systems. Its period of fluctuation is the longest of all known eclipsing binary stars, 9,885 days. This means it takes approximately 27 years for the star to change from maximum brightness to minimum brightness and back to maximum again! The main star of Epsilon Aurigae is relatively well known (1,900 light-years distant; 60,000 times brighter than the Sun), but its companion is still shrouded in mystery. To date, no trace of it has been detected despite the fact that it covers the gigantic main star, which has a diameter 100 times as large as the Sun's, for two years, the period of the variable star's minimum brightness. The unknown companion consequently has to be at least 1,500 times, or perhaps as much as 3,000 times, larger than the Sun. This would make it larger than any star we know. At this point astronomers consider it more likely, however, that this companion is a still young, hot star enveloped in a huge cloud of gas and dust out of which it was born and which is still so opaque that it obscures the light of the main star for over two years. The next obscuring of Epsilon Aurigae is expected to occur in the years 2009 to 2011.

Vega

This, the brightest star in the constellation Lyra and also the most conspicuous star of the Summer Triangle, is one of those stars that has nothing particularly unusual in its physical makeup but that is interesting for historical reasons.

In relation to the Sun, Vega is a distance of 25 light-years away, its diameter is three times as large, and its mass is four times as big. Vega's intrinsic luminosity is 45 times greater than that of the Sun.

Vega was the first star in the history of astronomy to be photographed. The photograph was taken on July 16, 1850 at the Harvard Observatory with an exposure time of 100 seconds and was made possible through the recently invented daguerreotype process. Vega's role as north star lies considerably farther back in the past. About 12,000 years ago the north celestial pole, which describes a very slow, large circle in the sky, was in the vicinity of Vega (see page 42). Polaris fulfills this function considerably better, however, for Vega was nine times farther from the pole than Polaris is.

Venus

The Roman goddess of love lent her name to the Sun's second closest planet. Venus approaches Earth closer than any other heavenly body except for the Moon and consequently shines very brightly in the sky. Its brilliant presence as the morning and the evening star (see pages 115–16) makes it a particularly striking object that always has seemed worthy of special veneration.

Modern astronomy has shown, however, that the conditions on the planet's surface do not resemble the image of the goddess of love in the remotest. Venus has a hellish landscape with surface temperatures of between 400 and 500°C (about 900°F) and an air pressure more than 90 times that on Earth. This kind of environment causes metals like tin and lead to melt, and the first space probes that were lowered through the planet's atmosphere survived these conditions only for a few hours.

The reason for the extreme heat on Venus is the planet's atmosphere. Unlike Earth's atmosphere, it is composed almost entirely—at least 98 percent—of carbon dioxide. Thus Venus represents an extreme example of the same greenhouse effect that is now causing general alarm on Earth. For it is the nature of carbon dioxide to let incoming sunlight pass through but trap heat radiating back in the same way that glass roofs do in greenhouses. It is this mechanism that heats up Venus's surface to such extreme temperatures.

There is of course no life on Venus, given the conditions that exist there. The planet's surface is known at this point only through detailed radar measurements, which show mountain ranges and plains. Venus's surface cannot be examined through a telescope because the planet's atmosphere is not transparent (see photo, page 148). Astronomers are gaining more insight from data sent back by the spacecraft Magellan launched toward Venus on May 4, 1989.

Basic Data of Venus

Distance to the Sun:	67.2 million miles (108.2 million km)
Period of revolution about the Sun:	224 days
Diameter:	7,517 miles (12,104 km)
Period of rotation:	243 days
Mass:	.81 times that of Earth

Index

Page references in **bold type** indicate color photos.

◁ *The Large Magellanic Cloud, the most gorgeous celestial object in the southern sky.*

Index

Credits

Photo credits:
Work collective ASTRO-FOTO, Alt, Brodkorb, Rihm, Rusche: Front cover, pages 5, 24, 25 (left), 147, 152, 153; TREUGESELL publisher: book spine, pages 10, 18, 25 (right), 129, 140; NASA/Bader Planetarium: pages 132, 139, 148; European Southern Observatory (ESO): pages 130, 136; NASA/J.P.L.: Front cover, pages 6, 7, 135, 144; SAO/US Naval Observatory: pages 17, 26, 52, 78; Doring: Front cover, page 109; Dreesen: page 110; Jim McCarthy: page 118.

Scientific advisor:
Rahlf Hansen, Dipl. Phys.
Position of planets: Hartwig Lüthen

Drawings:
Solar and lunar eclipses on flaps at front and back: Brian Sullivan; Star maps and diagrams: Wil Tirion and Brian Sullivan.

Illustrations of constellations:
From *Johannis Hevelii uranographia totum coelum stellatum,* Danzig, 1690 (reprint: Taschkent, 1978)

The author:
Joachim Ekrutt, Ph.D. has for many years been on the staff of the Hamburg Planetarium. He is also a contributor to the magazines *Stern* and *GEO* and the author of an illustrated reference volume on the Sun and of the Barron's Mini Fact Finder *Stars*.

The consulting editor:
Clint Hatchett has directed and produced programs at planetariums across the United States for many years, including the American Museum-Hayden Planetarium and the Christa McAuliffe Planetarium in New Hampshire. He has two published books on the night sky, one of which includes in-depth observing projects for the reader to try with simple homemade instruments.

The title of the German book is *Sterne und Planeten*

Translated from the German by Rita and Robert Kimber

All inquiries should be addressed to:
Barron's Educational Series, Inc.
250 Wireless Boulevard
Hauppauge, NY 11788
http://www.barronseduc.com

Library of Congress Catalog Card No. 99-35131

International Standard Book No. 0-7641-1310-0

Library of Congress Cataloging-in-Publication Data

Ekrutt, Joachim W.
[Sterne und Planeten. English]
Stars and planets : identifying them, learning about them, experiencing them / Joachim Ekrutt ; star maps and diagrams by Wil Tirion and Brian Sullivan ; maps of solar and lunar eclipses by Brian Sullivan ; contributing editor, Clint Hatchett ; [translated from the German by Rita and Robert Kimber]. — 2nd ed.
p. cm.
"With all important celestial events up to the year 2010 and with a lexicon of celestial bodies."
Includes index.
ISBN 0-7641-1310-0
1. Astronomy Observer's manuals. 2. Stars Observer's manuals. 3. Planets Observer's manuals. I. Hatchett, Clint. II. Title.
QB63.E3813 2000
523—dc21 99-35131
CIP

PRINTED IN HONG KONG
9 8 7 6 5 4 3 2 1

Planetariums and Observatories

United States
Grace H. Flandrau Planetarium
University of Arizona
Cherry and University Boulevard
Tucson, Arizona 85721

Planetarium
University of Arkansas
2801 South University Avenue
Little Rock, Arkansas 72204

Planetarium
Griffith Observatory
2800 E. Observatory Road
Los Angeles, California 90027

Reuben H. Fleet Space Theater and Science Center
Balboa Park, P.O. Box 33303
San Diego, California 92103

Alexander Morrison Planetarium
California Academy of Sciences
Golden Gate Park
San Francisco, California 94118

Charles C. Gates Planetarium
Museum of Natural History
2201 Colorado Blvd., City Park
Denver, Colorado 80205

Alexander Brest Planetarium
Museum of Science and History
1025 Gulf Life Drive
Jacksonville, Florida 32207

Jim Cherry Memorial Planetarium
Fernbank Science Center
156 Heaton Park Drive, N.E.
Atlanta, Georgia 30307

Faulkner Planetarium
Herrett Center
College of Southern Idaho
Twin Falls, Idaho 83303

The Adler Planetarium
1300 S. Lake Shore Drive
Chicago, Illinois 60605

SpaceQuest Planetarium
The Children's Museum
300 N. Meridian Street
Indianapolis, Indiana 46208

Arnim D. Hummel Planetarium
Eastern Kentucky University
Richmond, Kentucky 40475

Planetarium
Lafayette Natural History Museum
537 Girard Park Drive
Lafayette, Louisiana 79503

Davis Planetarium
Maryland Science Center
601 Light Street
Baltimore, Maryland 21230

Charles Hayden Planetarium
Boston Museum of Science
Science Park
Boston, Massachusetts 02114

Abrams Planetarium
Michigan State University
East Lansing, Michigan 48824

McDonnell Star Theater
St. Louis Science Center
5100 Clayton Road, Forest Park
St. Louis, Missouri 63110

American Museum Hayden Planetarium
Central Park West at 81st Street
New York, New York 10024

Morehead Planetarium
University of North Carolina-Chapel Hill
Morehead Building CB #3480
Chapel Hill, North Carolina 27599

Fels Planetarium
Franklin Institute Science Museum
20th and The Parkway
Philadelphia, Pennsylvania 19103

Henry Buhl, Jr., Planetarium and Observatory
Carnegie Science Center
One Allegheny Avenue
Pittsburgh, Pennsylvania 15212

The Cook Center Planetarium
3100 West Collin Avenue
Corsicana, Texas 75110

Universe Planetarium/Space Theater
Science Museum of Virginia
2500 West Broad Street
Richmond, Virginia 23220

Canada
Margret Zeidler Star Theatre
Edmonton Space Sciences Centre
11211 – 142 Street
Edmonton, Alberta
CANADA T5M 4A1

H.R. MacMillan Planetarium
British Columbia Space Sciences Society
Vanier Park, 1100 Chestnut Street
Vancouver, British Columbia
CANADA V6J 3J9

Dow Planetarium
1000 rue Saint-Jacques Ouest
Montreal, Quebec
CANADA, H3C 1G7

What do the maps of lunar eclipses on the following pages show?

Moon

Night side: Eclipse visible from beginning to end

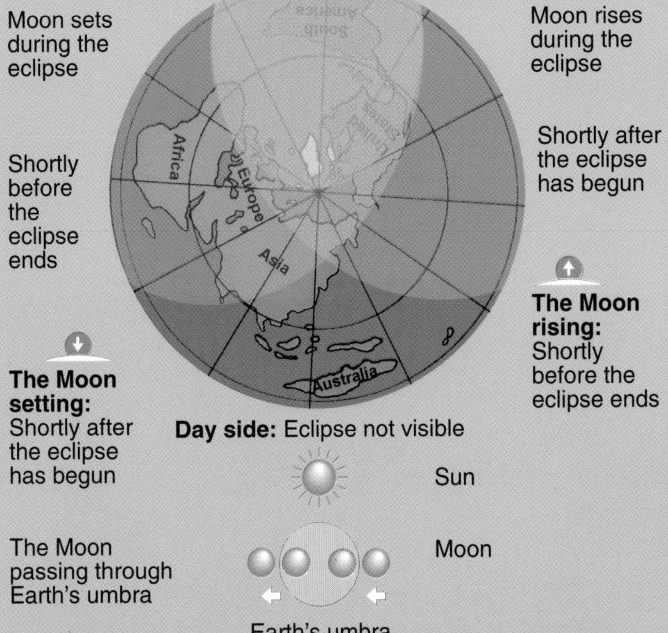

How Do I Use the Maps?

The maps on the foldout back flap show all the total lunar eclipses up to the year 2010. In a lunar eclipse, the Moon travels from west to east through Earth's circular umbra. The small drawings at the upper left show the exact times in EST of the eclipse. A lunar eclipse looks the same from all points on Earth where the Moon is above the horizon. The large maps of the globe, with the north pole at the center (the continents appear distorted), indicate where a given lunar eclipse can be seen. Where the Sun is shining at that time and it is day, the eclipse is invisible. On the night side, where it is full moon, one can see the eclipse for the full period. But because Earth rotates, there are two zones between the day side and the night side where the Moon is rising or setting while it is still moving through Earth's umbra. Inhabitants of these areas therefore witness either the end of the eclipse or its beginning. For more on lunar eclipses, see page 112.

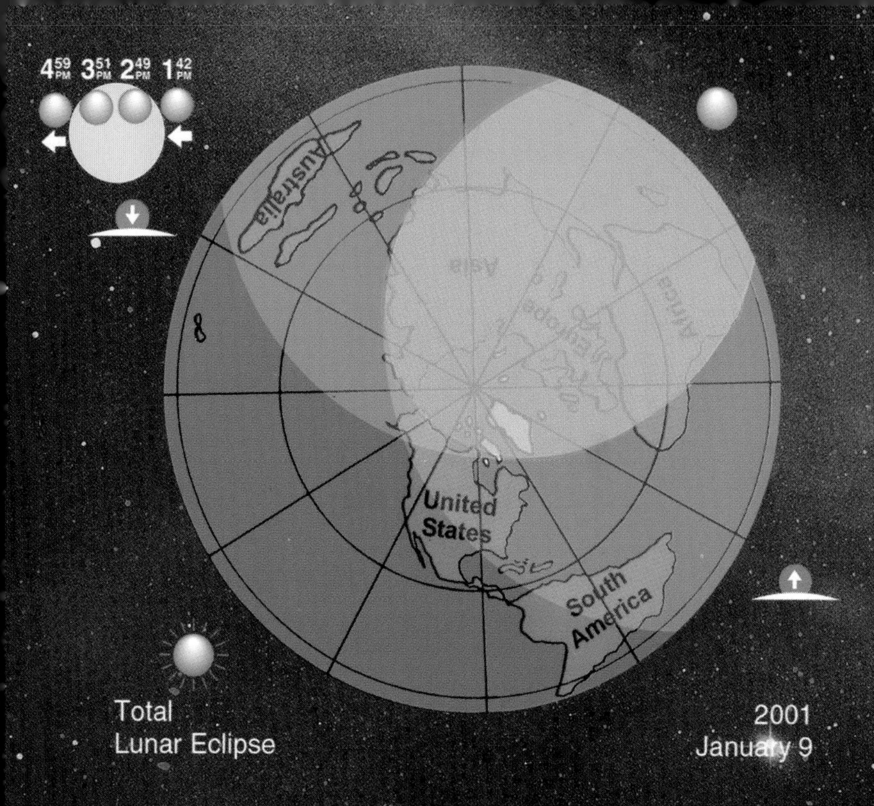
4:59 PM 3:51 PM 2:49 PM 1:42 PM
Australia
Asia
Europe
Africa
United States
South America
Total
Lunar Eclipse
2001
January 9

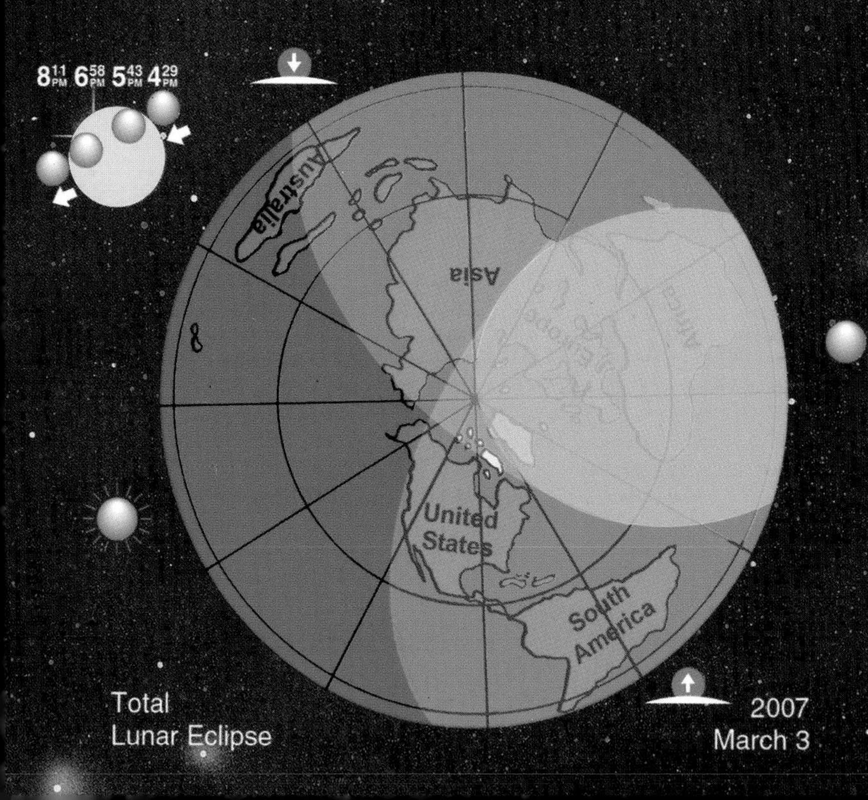
8:11 PM 6:58 PM 5:43 PM 4:29 PM
Australia
Asia
Europe
Africa
United States
South America
Total
Lunar Eclipse
2007
March 3